21st Century Estate Agency

Graham Norwood

2005

EG BOOKS

A di... RTC Limerick ...tion

3 9002 00035854 0

©Graham Norwood 2005

ISBN 0 7282 0459 2

Typeset in Palatino by Amy Boyle, Rochester, Kent
Printed by Bell & Bain Ltd., Glasgow

Contents

Preface

You are quite entitled to ask what, if any, qualifications I have for writing this book.

I am not an estate agent nor a developer; I am not a computer programmer nor an IT consultant. So why am I writing about how new technology can help estate agents to meet the challenges and opportunities of their industry in the 21st century?

I am a freelance journalist who writes exclusively about residential property for most of the UK national newspapers (chiefly *The Sunday Times*) and a few international ones, such as the *Wall Street Journal*, for some financial magazines, and for a handful of industry titles, such as *Estates Gazette*.

My interest in property is because, as a subject, it seems to have everything. There is constant intrigue and worry about the movement of house prices; it has long-term importance to the British economy; the public have a short-term fascination with it, wanting to know where the Beckhams and their celebrity peers live; and it has a genuine role to play in terms of people's well-being, whether it is buying a flat for a student child to live in, getting a retirement property for an ageing grandmother, or investing in a buy-to-let apartment to top up our woeful state pensions of the future.

There is an element of selfishness too. Over the past 20 years, I have owned nine homes in different parts of the UK, which means that I move house and become a client of estate agents about three times more frequently than the national average. Most moves have been for little more than fun or to cash in on equity gains, and I now have two investment properties. Therefore, as a punter, I am relatively well acquainted with the property market and estate agents.

As for new technology, I have an interest, but nothing more. Partly this is because I think it is fascinating in a sort of "gosh, look at what that can do" way, and partly because I know new technology helps me do my job better, so it may be safe to assume it can do the same for most others.

As with many estate agents around the country (and like almost all freelance journalists), I operate as a one-man band. Because of new technology, I work as a self-employed person from an office in the loft of my home in east Devon instead of having to be an employee based in a company office in London.

Around me I can see a PC and a glamorous large flat screen, although next to these are a beaten-up keyboard and over-clicked mouse showing their age. The PC has internet and e-mail connections. There is also a printer that can also be used as a fax, or as a photocopier, or even as a scanner to transfer hard-copy material onto a screen for integrating into documents.

There is a laptop computer that I use when I travel, which is fitted with one of those impressive wireless connections to the internet. This means that I can use the laptop as a word processor on the train and then send what I have written to some unsuspecting editor by e-mail from anywhere that has a wireless internet facility. My favourite is the Starbucks cafe in Regent's Street, central London, although there are many thousands of others across the UK. These days, new technology is everywhere.

Also, I have one of those strange, hand-held electronic devices known as a Personal Digital Assistant that I use as a diary and a contact book and a To Do list and even as a word processor (complete with a detachable fold-out keyboard) if I need to work somewhere and do not have the laptop with me.

There is a digital camera and a printer that allows me to churn out glossy old-fashioned snapshots from it, although most of the time I download the pictures on to the PC and e-mail them. There is a small tape recorder too, for when I interview people face-to-face and worry that my shorthand is not good enough — or if I anticipate they are the complaining or litigious type if they think I misquoted them.

Then there are the assorted bits and pieces that create a spaghetti junction on the floor. The lead for the mobile telephone charger, oversized batteries to power up the laptop, a telephone line (just the one, as broadband allows me to spend as long as I like on the internet and still receive and make telephone calls on the same line). There is a CD player in the room too, for when my latest utterings on the property market become too boring even for me and I resort to music.

This adds up to about £6000 of technology, possibly requiring £2000 of renewal each year. Is it worth it? Undoubtedly yes. It does not simply mean that I can do more work, more quickly and more easily than I would otherwise have done. It is worth it also because I can do some work I would never have even attempted prior to the invention of these devices, services and links.

Here is a true example. Several times a year, I am commissioned to write about the Asian luxury homes market for the North American edition of the *Wall Street Journal* (*WSJ*). Although I have worked in Asia during my 15 years as a BBC journalist, I now write these *WSJ* pieces sitting in my home office in south-west England using the internet, telephone calls and research books.

Only 10 years ago, I would probably have to have been in a London office, and would have relied on expensive telephone calls conducted at a time suitable for Asian office workers. I would have been happy doing that if I were employed by someone else who paid the telephone and overtime bills, but not as a self-employed journalist — it just would not have been worthwhile in cost terms.

But now technology is inexpensive and easy to use, and telephone calls are cheaper than ever. As a result, the volume of work I undertake is greater and my approach to how, when and where I do it are much more flexible than could have been the case a decade ago.

Therefore, I can receive information for those *WSJ* pieces by e-mailing my Asian contacts in *my* time and waiting for them to reply in *theirs*. The best agents across Asia have their properties displayed in detail on the Internet and if I need additional pictures, they are e-mailed to me in minutes. It is that simple.

Giving away trade secrets is unwise but I suspect editors who commission my work know that technology is responsible for much of it. Talking of which, my thanks go to Adam Tinworth of *Estates Gazette* magazine for suggesting that I write this book, to Alison Richards and Rebecca Chakraborty of Estates Gazette Books for being patient while I did the deed, and to all those who responded to my calls, e-mails and visits during my research.

Even after writing this, I confess to being no technology expert, as those who have seen me struggling with the timer of a video recorder can testify. But I do know how technology has made me run a better business, do more work, generate more income and leave more people convinced that I can produce what they want.

If a journalist can do that, so can any and every estate agent.

What Is 21st Century Estate Agency and Why Should We Care?

Estate agents should try to imagine this.

It is March 2010, the start of another sales season. Potential buyers are greeted in your high street office with a touch-screen PC enabling them to enter their property, personal and financial details as they sip their courtesy cappuccinos.

They touch another part of the screen and can see details of all the properties on sale in your area — not just apartments and houses whose owners have instructed you, but those on the books of nearby rivals too. By 2010, you may all be hooked up to a computer network and obliged to give customers details of every property on sale in the locality. Agents will have realised years ago that they could charge higher commissions by improving their service to the public through a US-style "multi-listing system" of all homes on sale in the vicinity.

The would-be buyers print out details and telephone you on your mobile to arrange a time to go on an accompanied viewing of their favourite three homes on sale. You are out on an inspection at the time but the remote link to your Personal Digital Assistant (PDA) hand-held PC automatically finds three convenient slots, so the appointments are made.

Meanwhile, back on your inspection visit, you draw up outside the potential vendor's home, straighten your tie and pick up your laser measurer and digital camera — you have had those for almost 10 years by now, haven't you? — before going in.

As you walk from room to room you enter the details on the hand-held PC and use an electronic stylus to draw up floor plans, comparing them with the downloaded details of the house from when it was sold last time, in 2007. You take an economical eight pictures of the house, mentally noting the best ones to use and deleting the redundant ones as you go.

1

You reach your suggested marketing price and discuss strategy with the vendors for a slow-burn or fast-action sales campaign. Just to impress and clinch the instruction, you ask to be excused while you pop back to the car for a moment.

Reaching into the back seat, you connect your hand-held PC to a laptop computer that already carries a pro-forma layout of the company letterhead. You transfer the room measurements and descriptions to the laptop, then download those digital pictures and slot them into place.

A few seconds for a read through while your right forefinger hovers over the print button and then click! The details appear on the back shelf portable printer, crisp and ready to show the impressed potential client.

You bounce back inside with a draft of their brochure. On the doorstep, you remember to ring the office from your mobile, to check that those same details have already been received via your laptop's wireless e-mail link. Your colleague at the office is just a few seconds away from putting them on the pending section of your internet website.

As the clients look impressed at the colour pictures of the house they are sitting in, you suggest they go upstairs to their PC to see how those same pictures would look on the internet. "They're already there in a private section, ready for you to check — that is, if you think we should get your instruction of course," you add coyly.

As the clients return, they pass the inevitable comment that just a few years ago estate agents were regarded as Luddites who shunned new technology and offered a slow and poor service to clients. Then, agents' fees were considered as overpriced but, although they are higher in 2012, they are considered better value. "How things have changed," your clients say, surprised but satisfied.

Small talk out of the way, they are ready to sign the contract. They just need a pen... a pen! Damn, the one thing you forgot. And, by 2012, perhaps the least important tool you will need as an estate agent.

Eventually, you find one and they sign. You remind them that on your website they can find details of the Home Information Pack that your firm is already preparing on their behalf, plus online quotes for your company's conveyancing service and a link to receive an immediate estimate on the price of removals on completion day.

You shake hands and retreat to your car, with one nagging doubt coming back to haunt you. Out comes your PDA and you type yourself an aide memoire: "Why have an office? I could run this business from my home and my car."

Fiction? For the moment, of course it is. Far-fetched? Actually no, not at all.

Each piece of technology mentioned above exists today and some enterprising estate agents in the US are already using it, even to the point of not operating from offices at all and instead carrying most of their equipment in their cars with the rest based in their homes.

But the real novelty is this — modern technology allows you to offer a fast, efficient service that could create a sea change in how the public in the UK perceive estate agents. Also, it could help you maintain and enhance your position in an increasingly competitive industry, and obtain higher commission fees if the industry as a whole changes.

But compare this with what happens today.

A world without technology

In September 2003, I sold my principal home and bought another, in line with about 1.3 million other people in that year. My experience was distinctly low-tech, not out of choice but because all the agents I consulted for valuations or instructions were small and they had not invested much in their offices or equipment.

The regional south-west estate agency chain I used to sell my house sent around the manager of their Exeter city centre branch. He measured rooms with a retractable rule, made notes in a note book, took two digital photographs of the outside ("one we'll use, one for luck," is how he put it) and proceeded to take six working days before producing a first draft of particulars for me to see.

Two more working days elapsed before the house appeared on the agency's own website or the internet portals (Assertahome, Rightmove and the like) to which his agency subscribed. It took a further week before the first advertisements appeared in the local press.

The property was valued at £365,000. Even at the modest UK level of estate agency commission (in this example it was 2%) was that the level of service or shining example of rapid technology that I deserved in return for the £7300 (plus VAT) that I would hand over if the sale happened?

The agent from which I bought a house was even worse.

He was a one-office agent in his 50s who ran the business alone, except when his 70-plus year-old mother stepped in for him while he attended viewings or pitches. The details of the £445,000 property I eventually bought were printed on poor-quality paper and containing out-of-focus photographs. They had a spelling error too.

His office did not have a computer. His front window contained details of properties for sale locally and in Florida (except his spelling condemned it to being Flarida). His system of recording buyers who completed the transaction and signed for keys was a notebook purchased from the local sub-post office. He had no system at all for recording offers made on properties, or his vendors' responses, and instead relied on his memory for passing on information.

Ironically, he advised me against having a full survey on the property I was buying (although it was over 300 years old) because he was sceptical of the professionalism of surveyors; as the transaction progressed in fits and starts, he blamed solicitors for delays as they, too, were not as good as they were cracked up to be.

Was all this just the local charm of a small-town estate agency? Perhaps it was really a step back to the good old days? It was both of these in appearance, certainly, but in what other business would such an unsophisticated, insecure set of systems be accepted?

Would people invest £445,000 through an Independent Financial Adviser (IFA) if he or she clearly did not have a computer in their office or use new technology to prepare routine paperwork (or even check the spelling)?

IFAs have improved their act as a result of being under stricter regulations and acting voluntarily to improve the poor reputation they had in the 1980s and early 1990s. Routinely, IFAs have laptop computers which they take to clients' homes; these can show projections of how proposed investments may change according to different financial scenarios, and can access the Internet to demonstrate real-time share prices or the latest City data from reputable impartial sources. This sort of approach is taken for small investments of under £5000, let alone ones of £450,000.

But my particular estate agent prided himself on his lack of training. He had run a hotel, been in the forces, travelled around the world. But did he have any skill in valuation? Or running a business? Or new technology? No, none of those skills existed he was proud to say, as he completed a manual diary of appointments.

Perhaps he was unusual, although certainly not unique. But, to me, as to every other client he had on his books, he represented the state of UK estate agency in 2003.

Although every industry has its technophobes, the relative lack of new technology currently used in residential estate agency in the UK may be down to how easy it is to set up in the business.

Estate agents today need no formal qualifications, have only two main pieces of legislation to take into account in their business, and require little capital to lease an office and establish a high street presence.

While there has been substantial investment in technology by many of the larger chains at the top of the industry — "top" being defined as those with the largest numbers of branches and the highest turnover within the major sectors of the residential market — there has hitherto been neither a requirement nor an incentive for smaller agents to invest in equipment to make their companies perform better.

This is despite the fact that the residential market has boomed in recent years. Prices have risen by an average of over 10% in 1999, 2000, 2001, 2002 and 2003, according to the Halifax house price index — in some years, by over 20%. Even in 2004, when the market became distinctly flat at the end of the year, price rises in most areas remained at well into double figures.

Added to this heady mix is the high tendency of Britons to move house (rising from an average of once every 7.4 years in 1994 to once every 6.1 years in 2001, according to the Office of National Statistics, although, as we shall see below this trend may be reversing at least slightly).

Put these factors together and you get a position where there has been plenty of business for every agent, big and small, modern and old-fashioned, qualified or not.

It's all change for the property market

The respected research department of property consultancy Savills, which is regarded as an industry leader and not influenced by the estate agency wing of the same company, believes that the residential market is in for a concerted period of lower turnover and smaller house rises.

It believes that a series of major upheavals in the residential market are at an end.

First, home ownership is no longer growing. The rapid expansion of ownership started by the Thatcherite revolution, through introducing public sector right-to-buy legislation and the denigration of the rental market, has reached its climax. Now there are far fewer people still wanting to buy who have so far failed to do so.

Second, the UK is now used to a long-term low interest rate environment. The general downward movement of interest levels of the past 10 years has been caused by greater stability in the economy,

Table 1.1 Estate Agents' League Table, 2004

Estate agency	Offices, January 2004	Change on January 2003
Countrywide Assured	865	+23
TEAM Association	550	−20
Connell	487	+335
Halifax	341	−19
Bradford & Bingley	307	+1
Your Move	291	−20
Spicerhaart	236	−7
Reeds Rains	133	−3
Arun Estates	120	No change
Kinleigh Folkard & Hayward	66	No change
Winkworth	58	+9
Hamptons International	55	+1
Chancellors	52	No change
Andrews	49	+8
Savills	46	No change
Bradleys	40	No change
Humberts	40	+7
Townends	40	No change
Knight Frank	38	No change
Cluttons	34	−4
Jackson Stops & Staff	34	+1
Keith Pattinson Limited	30	No change
Strutt & Parker	26	+1
Peter Alan	25	No change
The Venmore Partnership	21	+8
Acorn	20	+3
Burchell Edwards	20	−3
Dacre Son & Hartley	20	No change
Lane Fox	20	No change

Source: *Estate Agency News* magazine

discrete harmonisation with much of mainland Europe, and the 1997 switch of responsibility for rate movements away from politicians and to the Bank of England's monetary policy committee.

As the base rate tumbled between 1999 and 2001, so more people bought — either their own homes or buy-to-let investments or holiday properties. Now, Savills contends, we are used to a lower interest rate culture and the "rush" to buy has receded, especially as modest interest rate rises kicked in during 2004.

The peak turnover of residential property in Britain was in 2000, when some 2.2 million households were bought or sold, or both. By 2003, that reached 1.3 million, with 2004 unofficially estimated at 1.2 million and the Halifax bank predicting a further 10% reduction in 2005 as more owners settle for putting in extensions rather than incurring the stamp duty and associated costs of moving house.

Therefore, in blunt terms, there will be less business around for estate agents in the next few years.

Competition will inevitably become tougher between existing agents and any new entrants who moved to the residential arena — commercial operators, such as DTZ and King Sturge, have moved into the new homes sector in recent years, and top-end agencies, such as Knight Frank and Strutt & Parker (which were previously known almost exclusively for selling expensive country houses), have new homes divisions selling relatively modestly priced properties.

Those estate agents relying on old technology are likely to be the most vulnerable as our industry continues through the 21st century. Either they will be forced out or taken over.

The 30 largest estate agency businesses in the UK in January 2004 had 321 more offices in total than just a year earlier, mostly as a result of takeovers of smaller companies.

In late 2004, one of the largest takeovers in the UK residential industry's history happened, when Countrywide Assured took over Bradford & Bingley. The move meant that the enlarged Countrywide operation now has 1100 offices and about a 10% share of all residential transactions — more clout than any industry player ever before.

At the time of this book being published, ATIS Real Wetheralls is known to be considering submitting an offer to buy Cluttons. If the deal is done, the resultant firm will have over 1000 staff, mostly on the residential side, and it would become the eighth-largest agency in the UK. In turn, Cluttons admits to being in discussion with two smaller agents with a view to taking them over, so there is little let-up in the trend that big corporates keep getting bigger.

There are many reasons for the growth of larger firms while smaller agencies become increasingly vulnerable.

Financial services companies have moved into estate agency, either directly in the form of Halifax and the old Bradford & Bingley, or indirectly in the form of increased shares of ownership of agents by finance firms.

Many of these are importing sophisticated technology to modernise property sales and to produce customer-focused "one-stop shops" covering not only a new home but also a mortgage, insurance and removal services too.

Many hitherto-independent small firms have formed alliances. TEAM and Home Sale Network are the best-known and most successful examples of collections of free-standing firms that are operating collectively to produce administrative and advertising savings, and to produce a larger publicity machine. TEAM, for example, already uses advanced technology to communicate between its disparate branches.

In addition, the returns from residential sales have increased enormously in recent years, partly because of rising prices and partly because of a higher turnover of business. This phenomenon has drawn in much more ambitious, large-scale players who want to expand still further.

Gaining the techno edge

The Office of Fair Trading (OFT) says apartment and house sales in the UK in 1993 were valued at a total of £1550 million at 2002 prices; by 2002, that figure reached almost £3000 million. Since that time, some areas of the UK enjoyed price rises in 2003 of up to 30% and in 2004 of 25%, so the trend is likely to be accelerating.

Is your business equipped in such a way as to resist being swallowed up by larger players in the market? Or, if you are a major agency already, are you sufficiently equipped to compete in the future?

Agents often say that investment in technology is not worth the outlay simply because the public still do not embrace it themselves. Why spend money on technology when no one would use it, they ask.

In reality, the general public is far ahead of the estate agents.

In 1998, Britons sent only (only?) 21 million mobile telephone text messages every day; by late 2001 that went to 38 million and by summer 2004 the total was 61 million, according to the magazine *New Media Age*. (In China, in 2003, there were 15.6 billion text messages sent, by the way).

Every day, an average of 28.5 million internet pages are downloaded to mobile telephones in the UK — do not concern yourself over how inappropriate the small mobile telephone screen may be for assimilating the information carried on a typical internet page, just dwell on that number of 28.5 million. Do you know of any estate agent who is doing his or her utmost to tap into this market?

In January 2004, the last time a snapshot of UK internet usage was taken, some 59% of all homes owned a personal computer, with almost all of them (around 12.5 million households) having internet access. Almost 69% of small businesses also had internet access in the workplace. Of all of those, 3.2 million PCs had high-speed broadband connections evenly split between work and domestic connections.

A survey in September 2004 by Home Sale Network, the affinity group that links 750 independent estate agents around the UK, showed that 62% of its members found that the majority of homebuyers now consulted the internet before making contact with an agent (irrespective of whether that contact was in person, or by telephone or via e-mail). Once registered with an agent, the majority of homebuyers also downloaded property details, compared local prices and prepared a shortlist for potential viewing — some of the work hitherto done by the agents themselves.

"With more buyers using the internet to do their homework, they are able to give agents a much clearer brief of what they want. This enables the agents to focus their expertise and local knowledge in matching homebuyers to the properties best suited to their needs, with less time spent on requested visits to homes that are not appropriate," according to Home Sale Network's managing director, Richard Tucker.

If this is merely at the start of a technological revolution, what will the future be like? We can see a glimpse of it through the plans unveiled by the government to introduce Home Information Packs (HIPs) and other measures aimed at transforming the buying and selling process in the UK.

Whatever the rights and wrongs of HIPs, they are on their way. As we will discuss further in Chapter 6, "voluntary" HIPs will be allowed from July 2006 and the packs — complete with the new Home Condition Reports — will be mandatory for all home sales from January 2007.

In Denmark, where such packs are already mandatory and have been seen as models for a possible UK system by our own Office of the Deputy Prime Minister, estate agents have become the key providers. They do the significant majority of the work previously split, as in the UK, between agents, conveyancing solicitors, surveyors and sometimes other parties, such as buying agents.

The evidence from Denmark suggests that the large chains of estate agents were best placed to instigate training and introduce technology to liaise between different parties involved in the pack. As we will see in Chapter 6, the UK government's e-business initiative (an umbrella term to describe its drive to make all forms of local government and many private transactions switch to computerised formats accessed through the Internet) means that many components of future HIPs will be feasible only through the use of new technology.

Local authority planning and search departments, the gathering of title deed information, even the previous price history of the property coming on to the market will all be available electronically. The government is already creating vast data banks — through a system called the National Land Information Service (NLIS), again discussed in Chapter 6 — that will allow the creation of a pack to be straight-forward to any agent that has the appropriate technology and personal skill.

Will you be ready for that?

If not, you may miss out on another element of what happened in Denmark. That was the substantial increase in fees that all estate agents secured from individual clients as a result of being able to deliver a pack prior to sale, as well as undertaking the traditional responsibilities of marketing a property and finding a suitable buyer.

Indeed, Denmark's fees rose from a typical 2% commission in the late 1990s to 3.5% today, placing it much higher up the international league table.

What is perhaps most interesting from these figures is the sheer size of agents' commission in the US. Is it really a coincidence that agents in that country are the world's foremost users of new technology?

We will look at the US estate agency system and individual agents' use of new technology in that country in greater detail in Chapter 3. In the meantime, it is essential to see how US realtors have used new technology to simultaneously improve their service to customers, enhance their image, and boost their income.

At the core of realtors' use of new technology is the Multi-Listing System (MLS).

The MLS allows open access to property sales information across rival estate agencies. A seller registers with one realtor and is told that the house details will be made available to all other realtors subscribing to the MLS in that area. Likewise, a buyer who registers with one realtor is told the names of rival realtors in the MLS to prevent unnecessary duplicate registering.

So, if the MLS is widely used in an area, a buyer or seller has access to the majority of the local agents' services in one go without the absurd requirement of tramping around to dozens of agents' offices.

Table 1.2 International Comparisons of Average Estate Agency Fees Per Dwelling (2003 Prices)

Country or region	% fee assumed
Scotland	1.0
Ireland	1.5
Netherlands	1.6
Wales	2.0
England	2.0
South-East England	2.0
Melbourne (Australia)	2.5
Sydney (Australia)	2.5
Denmark	3.5
US (mid-west)	6.0
US (south)	6.0
US (north-east)	6.0
US (west)	6.0

Sources: Office of Fair Trading UK, Office of the Deputy Prime Minister UK, NVM, REI, Realkreditraadet, NAR

The biggest example of this in the US is the Multiple-Listing Service of North Illinois (MLSNI), which was set up in 1989. It has over 70,000 properties for sale on its database, and is estimated to carry 90% of the homes on sale in the area. It contains details of a further 1.75 million properties not currently for sale, plus over 500,000 photographs of properties — including interior, exterior and neighbourhood shots — and some four million property tax records relating to individual purchase taxes payable on properties, which are the US equivalents of council tax.

With little effort on the part of potential buyers they can look at comparable properties, sales histories, a rough visual condition of their chosen home, and an idea of how much it will cost to run.

The MLS system across much of the US is owned by 10 real estate associations (in some localities led by just one company, or on other occasions by a loose federation of agencies similar to the network set up in the UK by TEAM). In total, this represents a staggering 35,000 realtors — some working from offices but many operating from their homes.

Because of the widespread buy-in to the MLS by realtors, the information it contains is considerably more detailed and accurate than data from any similar UK system, such as Hometrack, which relies on data provided by small subsets of local estate agents who, allegedly, often forget to file their data or leave the task to junior staff.

When it comes to sales and completions in the US, the commission is shared among agents. For example, the person who introduces the buyer to a property will get a share; the agent on whose list the property is registered will get a proportion; if one or both of those agents works for a third-party broker, he or she too may get something; finally, it is often the case that the purchaser employs a buying agent who will also get something if involved in finding the property that is eventually purchased.

The larger distances and different methodology used by US property professionals is also reflected in where and how they base themselves.

The US National Association of Realtors says 80% of its members operate at least some of the time from a home office; a similar figure also operates some of the time from a conventional high street office; around 12% have what they call a "car office", which will have portable internet links, laptops and printers that will allow them to register properties on to the local MLS, as well as produce details and estimates on hard copy even though they may be in remote locations.

The critical difference is that, despite charging typical 6% commissions — three times those charged in the UK — realtors are not perceived as offering a poor service.

Unlikely as it may seem, a poll in 2004 by internet service provider America Online put realtors as the fifth most respected professionals in the US, partly because of their perceived widespread use of technology to improve the service to the public.

All of this is a far cry from estate agency in 2005 in the UK.

Modernising the UK estate agent

Most of those currently involved in modernising British estate agency and house moving (mainly a few government bodies and a plethora of firms selling software solutions) are loathe to comment frankly because they either need the co-operation of the estate agency industry, or they want to sell products to it. There are few statistics on the subject, but some broad indicators exist:

- Only 54% of estate agency branch principals can operate their office computer, according to research by the OFT.
- As of December 2004, broadband is available across 97% of the

UK but fewer than one in twenty estate agents offices have it, says pressure group Get Broadband Now.

- National Land Information Service (NLIS), at the heart of the government's e-conveyancing project, was launched in February 2001 but by July 2004 had only 4500 conveyancing solicitors registered to use it and only 61% of local authorities were able to handle electronic searches for property transactions.

Although many of our estate agencies in the UK are rightly proud of their past success, financial acumen and high esteem among individual customers, the overall image of the industry remains poor.

The number of complaints made to the Ombudsman for Estate Agents may have fallen during 2003 for the first time in the scheme's 14-year history, from 6462 in 2002 to 5356, but public scepticism at the findings of the OFT report in spring 2004 suggested this reversal had not heralded the arrival of widespread support for estate agents.

The Consumers' Association, with a vigorous publicity machine that helps set the agenda in much of the popular press, says it has identified "rogue agents who will flout the law" and says the industry is bedevilled by "unenforceable legislation [under which] it is very difficult to call them to account, allowing dodgy practices to flourish while law-abiding estate agents suffer a negative reputation, undeservedly".

Introducing a plethora of gizmos to your office and sprucing up a website will not counter such broad beliefs — a bad estate agent who buys new technology merely becomes a bad estate agent who can work a laptop. But how many agents have asked vendors if they want, or perhaps may even be willing to pay for, better quality details when they sell their home.

A ready-to-market package of professionally shot photos, a virtual tour, floor plans and quality laminated brochures printed on hard card costs between £300 and £500. "A few years ago, it would cost that just to hire a photographer and get prints done," claims Michael Doherty, the chairman of e-house, a software firm supplying the property industry.

"If you sell a middle-market property and you see one agent making photocopied details with a single badly taken snapshot and another using professional material, the latter has a good chance of getting the business even if the commission is higher," suggests Doherty.

"Vendors like professionally prepared particulars and will often pay for them — they recognise superior marketing can get a better price.

For the agent, the extra cost is recouped several times over by a small increase in their commission percentage," he says.

The price of ignoring all of this may be great to individual agents and the broader sales industry. Although the UK has avoided the high volume of "sell by owner" cases seen in the US, many lay people are identifying how easy it is to sell as home via the internet and other forms of new technology.

For example, vendor Christopher Bailey sold his home privately within four months.

"I needed to sell my property in north Yorkshire because of changes in the schools' catchment areas but I had 18 months to move so I wasn't in a desperate hurry. I did an internet search for "properties for sale in the north" and came across *www.countrylovers.co.uk*. They only wanted a tenner to advertise on the site so I decided to give it a go. I put my property on the market just before Christmas 2002 and was amazed at the number of responses — I had 27 enquiries by January," he says.

"After just a couple of viewings, I was made an offer and accepted. I got more enquiries after this and e-mailed people to say I'd stay in touch in case things went wrong. Previously, I'd had experience selling a house in Surrey — a lady made a serious offer but as the sale was going through we found out that her buyer was an American citizen who couldn't get a mortgage in the UK. By contrast, when I sold privately it was easier to deal with problems because I was in control of what was going on," according to Bailey.

"There was an issue about access to the property through some common land but the buyer and I reached an agreement between us; we took out an indemnity and each paid half. My property was on the borderline for stamp duty. I advertised it for £255,000 but accepted the offer of £245,000 — I was happy too as I'd saved so much on commission fees," he says.

There may be many risks in Bailey's activities and his story could easily have had an unhappy ending. But it didn't, and the number of individuals thinking along the same lines is growing. Is this a future that estate agents can afford to allow to happen?

There may be another advantage if the industry espouses more technology voluntarily. As the residential market flattens after the soaring prices of the late 1990s and the early 2000s, estate agents will have to work harder to keep their heads above water.

Cutting costs and creating a more modern impression through the use of new technology will encourage business, pure and simple. Read on to find out how.

Where We Are Now and How We Got Here

It is of no great surprise that estate agency in the UK has a relatively untechnical (some would say anti-technical) reputation. Although selling houses shares few characteristics with manufacturing or heavy industry, in the UK there is one common strand.

As with coal mining, steel manufacturing and underground trains, the UK created the estate agency industry before most of the rest of the world. Perhaps, as with other activities, we are now reaping the disbenefits of being an "early adaptor" and accepting practices that — although they worked well in centuries and decades past — have not encouraged continual reinvestment.

Mainstream UK estate agency as a profession is about two centuries old.

Prior to the late 18th century there were relatively few landowners and those that did exist belonged to a small and privileged class. In the Tudor period of the 1500s, for example, the population increased sizeably but, by 1600, the UK still had only 4.8 million inhabitants, of which just under 6% lived in urban towns defined as having 4000 or more residents. A mere 3% of the English population lived in London.

Much of the housing on the landed estates that dominated the countryside was rented out in some way, or was tied to the agricultural work of the tenants, and ownership passed from generation to generation within the same aristocratic families.

That ended with the industrial revolution, which shook the UK to its roots.

In the eighteenth century, a series of inventions transformed the manufacture of cotton in England and gave rise to a new mode or production — the factory system. During these years, other branches of industry effected comparable advances, and all these together, mutually reinforcing one another, made possible further gains on an ever-widening front. The abundance and variety of these innovations almost defy compilation, but they may be subsumed under three principles: the substitution of machines — rapid, regular, precise, tireless — for human skill and effort; the substitution of inanimate for animate sources of power, in particular, the introduction of engines for converting heat into work, thereby opening to man a new and almost unlimited supply of energy; the use of new and far more abundant raw materials, in particular, the substitution of mineral for vegetable or animal substances. These improvements constitute the Industrial Revolution.

(David Landes, *The Unbound Prometheus*, 1969)

Although there are few accurate population records, it appears that in 1750, roughly at the start of what most historians call the Industrial Revolution period, England and Wales had about 6.52 million inhabitants. This was 10% higher than just half a century earlier, but in the next 50 years there was a birth rate explosion. By 1800, the population soared to 16.35 million and then we were up and running.

Census and genealogy records suggest that by 1821 there were 20.98 million residents and by 1851 over 27.5 million. By 1881, there were 35 million and the dawn of the 20th century saw the population exceed 40 million. At that time, 77% of those people lived in urban areas.

Not only was this population increasing at a phenomenal rate, but also it was almost entirely composed of what some observers would later define as a working class. These were mostly families (usually large ones) that had to be housed near mines, foundries and plants.

As the 19th century industrial revolution continued much of the housing that was constructed was tied, so ownership remained in the hands of a relatively small number of people but as businesses changed hands so too did associated homes and the concept of private rent began to appear on a market-oriented basis. This meant the greater the demand, the higher the rent, and around particularly large cities and plants there were premiums attached to the costs of homes.

When private residential property began to be sold from one person to another, the earliest estate agents were members of the legal profession or were goods and chattels auctioneers who regarded transacting homes as a sideline to their main activity.

Lawyers in 17th century London were referred to as scriveners who "from being the clerks who wrote out or engrossed the deeds for the lawyers had become ... the expert conveyancing lawyers," according to

FML Thompson's *Chartered Surveyors: The Growth of a Profession*. Estate agents and formal surveyors did not exist as separate or respected professions at this time. Scriveners began keeping lists of estates for sale and of would-be purchasers who had been identified because they had put up notices in public areas requesting an interest in buying a home.

Other records of the time describe undertakers as some of the earliest estate agents. Alan Bailey's *How To Be An Estate Agent*, says 17th and 18th century undertakers, aware of those who had died, sometimes undertook to handle the dispatch of the property for the family as well as the dispatch of the body of the deceased. "Disposal of the family home and its contents seemed not far removed from the rest of the undertaker's services," Bailey writes.

"Property", "land" and "estate" are all loose terms here, for often even the most basic technology available at that time was not deployed to give potential buyers any true idea of size or condition. Plots of land were often described in terms of the numbers of animals that could graze or be reared on them; properties were "of a good size" (a term that still resonates with some more relaxed estate agents today, so perhaps things have changed even less than we think).

If there were room or field dimensions given, then they were sometimes established by rough twine stretched from one end to the other and then "measured" by the lawyer, undertaker or auctioneer stretching out the fingers of one hand and counting how many spans in the twine — and then converting that into feet and yards, or sometimes converted into areas appropriate to animal grazing.

There was reliance on good faith that the land was to be as fertile in the future as legend suggested it had been in the past, and that any buildings upon it were built in such a way as to last.

The process of a sale was similarly untechnical. Auctions were particularly popular as the real estate industry developed through the centuries, mainly because there were few records of comparable prices paid for similar properties. But, according to Thompson, "much house property and probably most landed property always changed hands by private treaty without ever reaching the auction room. Much property was put up for auction only to gauge its market value and was then withdrawn and sold privately. In pursuit of his business of disposing of a particular piece of property the auctioneer inevitably entered the world of private property negotiations and became an estate agent".

When the earliest formal auctions did take place, a candle was the closest to new technology that was involved. Samuel Pepys' diaries record a property auction encouraging bids over the duration of time it took for a candle to burn. Bidders in the crowd would shout a figure, slowly at first but then more competitively as the candle melted further. When the flame went out, the highest bidder had the house.

Handling the sale and purchase of residential property did not become anything like an industry in its own right until surveying and estate agency asserted themselves as professions in the second half of the 19th century.

The trade body that we know today as the Royal Institution of Chartered Surveyors started life in 1868 as the Institute of Surveyors, mainly for those surveying, valuing and managing buildings rather than selling them. But parallel with this, officially designated "selling agents" emerged with their own offices, usually near lawyers' offices in some of the better streets of cities. Some agents called on professional valuers to assist in pricing homes for some while some judged values for themselves, rather like today.

The volume of business for the estate agency business grew exponentially after the start of the 20th century.

Between 1900 and 1998, the housing stock of Great Britain grew from 7 million to 22 million, according to the Office of National Statistics and in the busiest ever year for new homes — 1968 — no fewer than 413,700 were built, about 60% of them in the burgeoning private sector.

This vast expansion of housing stock is mirrored, predictably enough, by the commensurate rise and rise of estate agents' offices.

It is now a major industry with around 12,000 separately owned estate agencies in England and Wales, with total fees from residential sales exceeding £2.5 billion per year. Only 6% of all properties sold in the UK do not involve an estate agent — they are passed either between friends or associates, or through private individuals acting as their own agents and advertising their homes in local or national newspapers and magazines.

Although this constitutes just a small proportion of the overall industry volume of business, it nonetheless represents about £25 million of lost income for agents every year.

The OFT's spring 2004 report into the industry analysed the latest available findings, concluding that:

We estimate estate agents' revenue from the sale of residential property in England and Wales in 2002 at over £2.5 billion. Between 1998 and 2002 there was an increase in the number of residential property transactions in England and Wales of nearly 19%, from 1.22 million to 1.45 million. Because property prices rose sharply over the period, the increase in the values of transactions was proportionately even greater. In 2002, total property transactions were some £185 billion, nearly 80% higher in real terms than in 1998.

After the rises of 2003 and 2004, the industry now must be even more lucrative. Yet, in our age of globalisation, corporate consolidation and brand homogenisation, it is probably a refreshing change to report that in this industry at least, the independents are still numerically as strong as the big chain operators despite the mergers of the past decade.

Corporates using technology to gain an advantage on independent agents

A 2002 survey by the Council of Mortgage Lenders (CML) and the 2003/04 investigation into the industry by the OFT concluded that "independent" operators (defined as those that were not tied by finances or management to corporate structures) accounted for almost 60% of estate agency offices in the UK.

Many independents are proud of this status and ensure they link up with other agents only for limited activities and economies of scale. This accounts for the growth of affinity groups set up to offer basic levels of administrative and market support, and to allow exchanges of properties — for example, offices of different estate agents in adjacent towns or neighbourhoods may share displays of properties in each other's patches to improve sales and the quality of service to clients.

Table 2.1 Estate Agents' Offices in Affinity Groups and Franchise Operations in 2004

National Association of Estate Agents' Homelink Network	832
Home Sale Network	742
The Guild of Professional Estate Agents	714
Movewithus	673
The TEAM Association	550
Mayfair Office	342
The London Office	184
Legal & General Franchising	118

Source: Council of Mortgage Lenders/Office of Fair Trading

Despite the trends towards aggregation, the big corporate estate agents, such as those owned by Your Move, Sequence, Halifax and Countrywide, still account for only about 40% of offices. To the public, the corporates' dominance may appear even less powerful because many of these big companies operate through agents and branches trading under well-established local names in different parts of the country. Countrywide has stated its intention to move further in this direction throughout 2005 by re-branding its newly purchased Bradford & Bingley offices with historic local agency names if possible (even if, in some cases, the original owners of such names went out of business some time ago).

But this image of an industry still dominated by small companies, although superficially warming to those keen on nostalgia, produces inverse results when the use of new technology is assessed. The corporates, unsurprisingly, have invested significantly in hardware, software and training — but the small firms have not.

The CML survey, which includes almost all of the professional players in the house transaction chain (that is, estate agents, conveyancing lawyers, mortgage brokers and surveyors) suggests that, although by 2002 estate agents were beginning to dip their toes in the new technology water, corporates were far ahead of independents.

The big players were in a particularly strong position when it came to using more advanced forms of new technology, such as:

- Virtual tours of property or video streaming — this is when a potential buyer with a standard home or office PC and internet connection can "travel" from room to room to get a detailed view of the property. Although most of the 40 major corporates surveyed by the CML offered virtual tours, the facility was available on very few websites (and even then on few properties on those websites). This was because of the cost of the software and preparation, and therefore it was used mainly on "the better instructions" — that is, the more expensive homes.

- A few of the corporates were experimenting with sending text messages to their buyer clients' mobile telephones — this was usually to inform them of new properties that had come to the market.

- E-mail was also becoming more important — some direct conveyancing firms (that is, those centrally operated from call centres and handling most customers through electronic communications rather than through personal meetings) suggested that up to 50% of their contact with customers came via e-mail.

But anything beyond these relatively long-established forms of new technology were effectively too exotic for most property professionals. The CML survey concluded that, for example, almost no one considered using Personal Digital Assistants (PDAs), also known as "palm" devices because of their small size, even though these could be used in the collection of information with software-enabling estate agents to produce floor plans and other standard particulars while onsite. This data could then be uploaded to a computer for integration into a brochure and subsequently for printing. However, the number of agents who even had PDAs was so small it did not feature in the CML's conclusions.

The one area where independents were almost on a par with the corporates was in the basic use of the internet.

Yet even here the corporates retained many advantages. For example, they had the funds to make elaborate websites with better graphics. Then, because of their multi-office network and subsequent economies of scale, they could offer a nationwide coverage of the market, allowing people in one part of the country to look at properties on the market in another area. The corporates could also use their larger bargaining power to forge alliances with associated services offered on their websites, such as links to mortgage providers, conveyancing solicitors and removal firms.

Even so, the CML found the independents trying very hard to compete:

- Without exception, all the independents and corporate estate agents questioned in the survey had their own websites.
- The majority of independents and corporates were associated with property portals (that is, those sites aggregating the property portfolios of many agents) such as Rightmove, Assertahome, Fish4Homes or Primelocation.com. These helped give a wider geographical coverage than any stand-alone agency website could offer.
- Many independents belonged to at least one affinity group, such as Home Sale Network or the TEAM Association; this allowed economies of scale in terms of money spent on website creation, branding, and, in some cases, the formation of a small version of US-style "multi-list" displays of properties on sale in the same locality, even from competing firms.
- A small number of independents also subscribed to technology-motivated affinity groups, such as The London Office that

"hosted" property pages (meaning that they serviced websites) for smaller estate agencies.

Even so, the corporates had immense advantages and this was shown nowhere more than in technology investment. Connells, for example, has a string of training courses in how to operate and exploit new technology to win new business.

Is the lack of regulation to blame?

A partial explanation of why small or middle-sized estate agency businesses do not invest in new technology is that the entire industry is relatively unregulated in legal terms and, therefore, actively encourages small firms to enter the field with few assets. Whether they can stay competitive and even keep operating against increasingly technological competitors as the corporates is another question.

Estate agents have only two main pieces of legislation created solely to regulate their business.

The first is the Estate Agency Act, which was introduced on 4 April 1979 in the dying days of prime minister James Callaghan's Labour government but did not come into force until 3 May 1982, when Margaret Thatcher's Conservative administration dominated British politics.

The Estate Agents Act 1979 was intended to regulate the work of estate agents so that both buyers and sellers were treated fairly and honestly. It covers activities such as how buyers' offers should be handled, the information agents must provide to clients at the start of an instruction and throughout its lifespan, and the way in which certain terms must be explained if used in estate agency contracts or publicity.

The act imposes a "negative licensing" regime in which an agent shown to have breached certain provisions of the act, and/or to have been convicted of certain criminal offences, may be banned from continuing in the profession.

This Act was effectively the culmination of 14 abortive previous attempts in the House of Commons since the start of World War One to introduce formal registers of professional house sales agencies as well as architects, surveyors and valuers. Those 14 attempts were each defeated after lobbying from organisations, such as the Incorporated Society of Valuers and Auctioneers and then latterly the Royal Institution of Chartered Surveyors and National Association of Estate

Agents; their stances ranged from a wariness about registration to outright opposition, and there was no shortage of MPs to express their reservations during debates.

It is no surprise that even in 1979, the then Labour government — even though it was perhaps the last administration that tried to exert substantial state controls over any industry dominated by private companies — nonetheless shied away from demanding the outright registration of estate agents. Instead, it used the act to give some protection to members of the public against any estate agent found to be dishonest or incompetent.

As a result, the Act, amended and watered down after lengthy and acrimonious debate especially at Commons committee level, merely asked for estate agents to demonstrate proper accounting systems when handling deposits (a few well-publicised law cases at the time involved miscreant agents who allegedly ran off with deposits).

The Act specifically maintained the unregulated nature of the business by stating that anyone could practise as an estate agent until (in the eyes of what has now become the OFT) he or she showed an "unfitness to practice". To emphasise the light regulatory touch, the government specifically excluded residential developers and other property professionals, such as planners, surveyors and valuers, from the legislation.

The second of the regulatory laws applying to agents came in 1991 after several years during which the Thatcher government sought to include all the formal and informal groups connected with property transactions into various trades descriptions legislation of the day. Thus was born the Property Misdescriptions Act, which eventually became law in April 1993.

The Act, in its own words, "applies to statements made in the course of estate agency or property development business for the purpose of disposing or acquiring an interest in land." It thus covers both residential and commercial property, but developers are subject to its conditions only when they offer for sale "an interest in land which consists of, or includes, a building constructed or renovated as part of that business". Therefore, a developer seeking to dispose of land or buildings surplus to his requirements which he had not developed or redeveloped himself would not be subject to the provisions of the Act.

Of course, other laws and guidelines from the likes of the OFT impinge on the activities of estate agents but we can see that, generally, it is a lightly regulated industry. The two main laws are interesting because of what they say agents should not do, rather than specifying what they should do by way of providing minimum levels of service to clients or specifying minimum levels of competency of those entering the industry.

Perhaps, as a result of this, the estate agency industry in the UK is often regarded by the public as having a poor reputation, especially when high levels of publicity are given to complaints and problems caused by the proverbial few bad apples. Yet, whether we think it justified or not, there is plenty of statistical evidence to suggest estate agency remains heavily associated with high levels of customer dissatisfaction:

- The OFT report of spring 2004 said 21% of sellers and 23% of buyers believed they had experienced "a serious problem" with at least one estate agent during each of their property transactions, and were able to identify specific shortcomings to support their view.
- In a minority of cases these included serious shortcomings such as apparent failures to pass on offers; suggestions that a client might be more successful in finding an appropriate house if they used the financial services offered by the agent; and failure by the agent to declare a personal interest.

The drive for more reliable and potentially more bureaucratic management information from each estate agent's office is made clear in the OFT's report into the industry in spring 2004. The findings conclude:

> Because there are no requirements for agents to keep administrative records of offers received and passed on to vendors, it is very difficult for the Office of Fair Trading and the Trading Standards Service [operated locally by trading standards departments in local councils] to substantiate complaints relating to matters of this kind. A more robust method of record keeping should be required so that a clear audit trail is available to the OFT and the TSS. This would help deter malpractice in the first place.

In an age and a society in which the consumer is king, the likelihood is that there will be further regulations introduced on agents' performance to follow up the successive Blair governments' concept of Home Information Packs (HIPs). After all, HIPs — in the government's eyes, at least — will simplify and hasten the sales process and should give more information and influence to buyers in particular.

Stricter money laundering regulations have already been introduced and few would bet against future requirements being introduced for better records of property details, viewings and offers made. There may also be a need for continually updated "log books" on the state and maintenance of homes, of the kind kept by many owners of properties in the US and handed over to successive owners to maintain.

Estate agency has come a long way and legislative and consumer spotlights are now on the industry more than ever before. This makes it even more important that those working on the front line have the right technological tools for their job. In the next chapter we see exactly how agents elsewhere use those tools.

A Different World — How Estate Agents Elsewhere Use Technology

By now, any self-respecting estate agent will be justifiably accusing me of taking the populist course. I criticise estate agents in the UK for allegedly being old fashioned and reluctant to change, yet it is not really any different elsewhere.

In some places at least, things are different and, in many people's opinions, better.

There has been a spate of recent studies comparing the UK transaction process with those of other countries. Not many show the UK in a flattering light although this is by no means the fault of our estate agents. For example, look at the tables below.

Scotland is treated separately to England and Wales because of the different house-buying legislation and because most transactions are handled wholly by solicitors.

Table 3.1 International Comparison of Average Length of Time, From Offer to Completion of Sale

Country	Offer to agreed terms	Completion	Total weeks
Australia	2	6	8
Denmark	2	4	6
Ireland	2	2–3	4–5
Netherlands	1–3	3	4–6
Scotland	–	–	8–9
US	1	6	7
England and Wales	8	4	12

Source: Office of Fair Trading, 2004

What this does not take into account is any difference made by the pilot "single survey scheme", which was introduced in Edinburgh, Glasgow, Dundee and Inverness in autumn 2004. Under this scheme, sellers are encouraged to arrange and pay for a single comprehensive survey to be given to potential buyers when a home goes on sale.

Until now Scottish buyers, along with those across the UK, have had to pay for their own surveys during the purchasing process. Because Scottish estate agents often encourage competitive bidding for homes, this means some homes have multiple surveys from rival buyers, sometimes even conducted on the same day.

Therefore, sellers suffer frequent visits from surveyors acting for a large band of potential buyers, while individual vendors suffer because each may pay for a survey although they know that (possibly) several such surveys already exist — perhaps even done by the same surveyor but for different would-be buyers.

If single surveys speed up the transaction process, Scotland may shoot up the league table but this will not change the findings of other OFT research in 2004. Look at the tables below.

Table 3.2 International Comparison of Membership of Professional Bodies

Australia	Up to 90%
Denmark	95%
Ireland	40%
Netherlands	88%
Scotland	100% (solicitors); 25% or less (estate agents)
US	75%
England and Wales	25% or less

Source: Office of Fair Trading, 2004

Table 3.3 International Comparison of Compulsory Registration of Agents

Australia	Yes
Denmark	Yes
Ireland	Yes
Netherlands	Yes
Scotland	Yes (solicitors); No (estate agents)
US	Yes
England and Wales	No

Source: Office of Fair Trading, 2004

Table 3.3 International Comparison of Estate Agents' Educational Requirements

Australia	40-hour course and examination
Denmark	Compulsory real estate higher education and examination
Ireland	None
Netherlands	Voluntary real estate higher education and examination
Scotland	Legal training for solicitors; none for estate agents
US	30–90 hours of training, depending on state laws
England and Wales	None

Source: Office of Fair Trading, 2004

Only in terms of simple pounds and pence does Great Britain come out "better" than almost any other country studied — although whether agents comfortably accept the relatively low fees they charge in this country is another question.

Table 3.4 International Comparison of Average Agency Fee Per Dwelling (2003 Prices)

Country or region	£ equivalent
Scotland	727
Wales	1667
Ireland	1890
Netherlands	2183
England (not south-east)	2679
Melbourne (Australia)	3012
South-east England	3526
Sydney (Australia)	3568
Denmark	4953
US (mid-west)	5095
US (south)	5518
US average	5922
US (north-east)	6146
US (west coast)	8009

Source: Office of Fair Trading, 2004

The UK compares quite starkly with other countries on the subject of qualifications required to become an agent or even take the more modest role of a sales consultant in an agency. Many countries have strict pre-entry requirements and licences — there are four examples below, all supplied by the UK Consumers' Association, where minimum standards have to be met before a licence is granted.

Country: Australia (New South Wales state)

Sales staff licence: individuals must pass a 40-hour course held over five successive days and managed by the Real Estate Institute of New South Wales (REINSW). Modules include an introduction to real estate, the use of technology, and consumer protection.

Estate agent licence requirements: individuals must hold a sales staff licence and have passed a separate 30-day REINSW course comprising 16 compulsory units, including how to communicate effectively with clients.

Country: Canada (Ontario region)

Sales staff licence: applicants must undergo 200 hours of training, including a semi-legal "articling" course and learning modules on marketing and promotion, valuation, estate agency principles and property law.

Estate agent licence requirements: individuals must win a sales staff licence and then obtain passes in exams and assessment of modules on strategic management, business plans and IT.

Country: Hong Kong

Sales staff licence: there are 30 hours of training in estate agency principles, law, land registration and consumer protection.

Estate agent licence requirements: applicants must undergo the 30-hour sales training plus 50 hours on property valuation, technology, management and business planning.

Country: Denmark

Sales staff licence: no training required.

Estate agent licence requirements: individuals must undergo two years of academic study and two years of practical experience working as an apprentice at a licenced estate agency. The final exams cover technological skills, law, valuation and land issues.

Of course, many UK agents undertake excellent on-the-job training and encourage staff to upgrade their knowledge throughout their careers.

For example Northfields, a three-office sales and lettings agency in west London, is unusual in winning an Investor In People award for its staff development despite being very small. It encourages staff to take training courses run by the National Association of Estate Agents (NAEA) and Association of Residential Lettings Agents, operates an in-house mentor scheme, where a more experienced staff member helps a newcomer, and even contributes financially towards external qualifications (in one case, paying half of an employee's foreign language course).

But for every Northfields there are others that rely on just getting through.

So where does technology come in to all this? Calculating the skills of updating a website or the application of improved customer care skills does not lend itself to a table or to straightforward international comparisons. But the case studies below may give food for thought on how the estate agency cultures of some countries link technology to higher standards and skills. How would we rank in the UK compared with these?

Case study 1: the US — the multi-listing system, technology and qualifications ... a potent mix

Imagine a place where residential estate agents use the word "ethics" to describe their business, where there is no public investigation into their practices, and where rival agents share property sales and commission in order to better serve the public.

Unlikely as it seems, that place is the US. Just to prove it, a recent poll by America Online put "realtors" — as US residential agents are called — as the fifth most respected professionals in the country.

It is not only thousands of miles that separate the US and the UK countries, but light years in the way agents conduct themselves.

The way in which US "realtors" operate has been looked at the Office of the Deputy Prime Minister (ODPM) prior to its much-vaunted unveiling of the HIP legislation, and also by the OFT during its 2004 survey.

One area looked at by both investigations was registration. The US National Association of Realtors (NAR) is the world's largest professional association, with 1.59 million licensed real estate agents, of which around 600,000 members are inactive (that is, the licensees have been estate agents but have now left the profession through retirement or career change) while 950,000 are current, full-time practitioners. The NAR claims that only around 8% of practising real estate agents are not members.

"The secret [of our success] is our adherence to a strict code of ethics and the commitment of members to go through life-long learning to understand real estate laws and finances, valuation issues and, of course, their neighbourhoods," according to NAR spokesman Walt Molony.

But that code — which at a hefty 6,400 words long, appears in full on the NAR website at *www.realtor.com* — is only the figurehead on top of a raft of practical measures which, in the NAR's views, make US agents better respected than those in other countries.

"At the centre of our operating methods are licensing and qualification. Almost every individual state sets out local requirements, which have to be met before someone can deal in real estate," explains Molony, who believes an awareness of how to use technology is of benefit to agents and their clients alike.

This usually involves the trainee attending evening or day classes in real estate law and practice, including valuation and market report writing using online technology, and concluding with a three-hour exam, which must be passed.

The 2003 paper from the South Dakota Real Estate Examining Board, for example, covers the different types of real estate agent; the legality and appropriateness of different sources and types of home loan; contracts; "debt-income" ratios; damages clauses in purchase contracts; financial leverage; estate and inheritance tax, divorce settlements, capital liquidation and merger, and loan foreclosures; and the investment potential of property.

There is also training in how the multi-listing system operates and in the practicalities of software application.

"In most states, people qualify to become a realtor by getting an agent's licence; they will often work for a broker who is better qualified still. He can be in charge of a realtor office and is ultimately responsible for the activities of the licensed agents operating from his office. The broker can also join a board of realtors, which helps set and assess real estate professional standards in a state," explains Molony.

Once qualified, US realtors operate in a highly commercial but collegiate environment where technology takes centre stage. At its centre is the Multi-Listing System (MLS), the most overt use of technology to give what realtors believe to be a real commercial advantage.

The MLS allows open access to property sales information across rival estate agencies. A seller registers with one realtor and is told that the house details will be made available to all other realtors subscribing to the MLS in that area. Likewise, a buyer who registers with one realtor is told the names of rival realtors in the MLS to prevent unnecessary duplicate registering.

As such, if the MLS is dominant in an area (as is the case in many large cities), a buyer or seller gains access to the majority of the local agents' services in one go.

An example of MLS listings can be found on *www.chicagometro realestate.com* and *www.realtor.com/chicago*. The listings do not only give buyers a view of most of the properties available, but also:

- the average sale price of recent property transactions in a neighbourhood
- the average number of days each property was on the market before completion
- the number of sales over fixed periods, to give an indication of how active the market is
- the percentage change in sale prices over any period dating to 1989
- a breakdown by house type (detached, condiminium, etc) of the total housing stock, of those on the market, by average price, location or speed of sale.

Because of the wide support given to this system by realtors, this information is considerably more detailed and accurate than information from some similar UK systems, which rely on data provided by small numbers of local estate agents.

"The only negative is the one associated with all online or internet databases. If you don't put good information in, you won't get good information out. So some agents that are unfamiliar with properties or do not update the data [frequently and quickly] can inadvertently produce poor information. But otherwise the advantages are huge," says Mark B Weiss, a Chicago realtor who is one of the contributors to the MLS, has written four books on industry topics, and lectures trainee realtors in California.

"The standard commission on most properties is around 5%–6%," says the NAR's Walt Molony.

He explains how that is shared: "It can be split amongst up to four agents. For example, the person who introduces the buyer to a property will get a share; the agent on whose list the property is registered will get a share; if one or both of those agents works for a broker, he or they may get something too; finally, it is now often the case that the buyer employs a buying agent, who will also get something if involved in finding the property that is eventually purchased."

Indeed, a survey of several thousands of house sales in the US showed that 63% of buyers now employ a "property finder". Unlike in the UK, where finders charge an additional fee and are associated with expensive properties, the US equivalents are used across the whole market and take their fee out of the pre-agreed commission.

Mark Weiss says buying agents are most useful for those seeking the cheapest properties.

"If you've bought several homes in the past and you're comfortably off, you're unlikely to make mistakes with a purchase. You know what to do and when to do it. First-time buyers, who are critical to our real estate economy, are much more likely to need a buying agent," says Weiss.

In addition, because estate agency fees are high in the US, around 1 property in every 20 is sold by the owner simply by putting a sign in the garden or letting community contacts know it is on the market (the UK figure for private sales is around 1 property in every 160). In the US, buying agents are likely to know of these private sales and thus can help the inexperienced or out-of-town purchaser.

The larger distances and different methodology used by US property professionals is also reflected in where and how they base themselves.

NAR surveys show that 80% of realtors operate at least some of the time (two days or more per week) from a home office; a similar proportion also operate some of the time from a conventional high street office. "As we have discussed, around 12% have what they call a "car office", which has portable internet links, laptops and printers that allow them to register properties on to the local MLS, as well as produce details and estimates on hard copy even though they may be in remote locations.

That scenario in Chapter 1, where draft details are bought to a vendor before the visiting estate agent returns to the office, is already coming true in the US.

In addition to internet sites, realtors' shop windows, press advertisements and free local property papers, the US realtor has a further marketing ploy that is used in the UK only by Hamptons International, Connells and a few independent agents — that is, the "open house" marketing technique.

"These tend to be on Sundays and we invite into a vendor's house the local people, registered buyers or anyone else who may be interested. The seller is away, so the realtor conducts proceedings. People can see what's going on and it is a useful way of buying agents introducing themselves to potential clients," admits Weiss.

In many states, it is obligatory for homeowners to keep "logbooks" of receipts, guarantees and certificates concerning planning permission and work on property; in addition to the more formal title deeds, these are usually presented — free of charge — to buyers when they consider making an offer on a property for sale.

Most US realtors consider this a sensible process to speed up sales and give buyers the maximum information on which to base a purchase. If this sounds like a version of the HIP, it is — the logbook was looked at by the ODPM in its research before floating the HIP idea.

Of course, not everything is rosy in the US garden. There are significant problems faced by the US real estate industry, as in the UK, but these tend to be market-oriented and not based on public image.

For example, the NAR is waging a campaign of opposition to a proposal pursued during President George W Bush's first and second administrations to allow banks to act as real estate brokers and set up property management businesses. Individual realtors have had to change the way they advertise properties on the internet to avoid clashes with anti-trust legislation, which is also causing concern. As in the UK, there is concern in the industry that first-time buyers may be excluded from purchasing by fast-rising prices, thus stoking up problems elsewhere in the market.

But the critical difference between the two countries remains clear. It is that despite charging an average commission of 6% — three times that levied in the UK — realtors in the US are not perceived as offering a poor service.

"My American colleagues just cannot believe that we do not have licensed estate agents," says Peter Bolton King, the new chief executive of the UK's NAEA.

"In the US, minimum standards and state licensing are the norm and this is something the NAEA has been asking the UK government for many years," he says. Yet the NAEA remains resolutely opposed to a multi-listing service.

By way of context, this is how the US buying process works and how a buyer perceives a benefit from the MLS.

Once a buyer decides on a property, he makes an offer — but this is a much more significant step than it is under the UK system. If accepted (and it may take some negotiation to reach this stage), this offer becomes a binding sales contract (known in different parts of the

US as a purchase agreement, an earnest money agreement or a deposit receipt). This serves as a blueprint for the final sale and includes:

- address and legal description of the property
- sale price
- terms (for example, all cash or subject to obtaining a mortgage)
- seller's promise to provide "clear title"
- target date for closing the sale (equivalent to completion in the UK)
- the amount of deposit (returned if the offer is rejected, or kept if the sale proceeds or the buyer backs out without good reason)
- an agreement on how utility bills will be settled during overlap of seller/buyer ownership of property
- an agreement on who pays for the survey, environmental inspections, searches, etc
- a UK-style fixtures and fittings list
- a provision that the buyer may make a last-minute walk-through inspection of the property just before "closing" the deal
- a time limit (normally short, often just two weeks) after which the offer expires.

If the seller agrees with most of the contents of the offer but has some disagreements — which could be on anything from the price, down to the contents of the fixtures and fittings list — then a "counter offer" can be proposed. The buyer then accepts or rejects the latest proposal, or makes a further counter offer. The document finally becomes a binding contract when one party signs an unconditional acceptance of the other side's proposal.

A buyer can withdraw an offer up to the time it is accepted although any withdrawal after that point leaves the buyer open to costly litigation and/or the loss of a deposit.

At "closing", documents that transfer legal ownership are signed and closing costs — realtors' fee, attorney's charges, taxes, survey cost and search fees — are paid at that time by the seller or the buyer, whichever is specified in the agreed contract. In reality, this is almost always the seller and these fees can often add up to 8%–10% of the cost of the property.

American-born Anne Graves, a public relations consultant working in the UK, has bought houses in Massachusetts in the US and Surrey in the UK. She comes down firmly in favour of the US system.

"In the US, realtors are more likely to be women than men and they are usually active in the local community. They belong to the country clubs, golf clubs, have kids in the local schools — so they know who is selling and why," says Graves.

"I had identified the town I wanted to live in [in Massachusetts] and had contacted several realtors to set up appointments. Several of the properties were on with all the agents through the multi-listing service. I was always accompanied on visits, so it was possible to have a really good look around and to discuss the property with the agent on site," she says.

"Each realtor knew the area thoroughly and gave me useful advice on schools, clubs, shops, local handyman, etc. And when we moved in we received a huge basket of homemade bread, jams, wine and fruit from the realtor — it was really lovely."

Graves bought a four-bedroom detached house in the US; in the UK, she bought a five-bedroom house in a village near Guildford.

"It was really hard work — virtually a full-time job. Each of the agents had different properties so you had to go to each to see the range available in a given area. I had to find the places and visit them on my own, with a tour by the owner, which inhibited looking round and asking questions," she explains.

Case study 2: the US — e-mail newsletters

The only UK estate agency that regularly sends out an e-mail newsletter to potential and current clients is Savills. In the US, it is relatively common, with the NAR selling individual agents' software packages.

E-mail newsletters work like this. A monthly e-newsletter is designed by the NAR and sent to individual members. It includes new consumer-oriented articles about property, including current market conditions and interest rates, plus general home ownership lifestyle articles. All the editorial is written by professional journalists hired by the NAR.

The software allows the individual agency offices to add their own highly specific local articles, their own branding and contact information. The NAR sees clear advantages in this approach — it allows agents to keep in touch with clients without the intrusiveness of personal contact, hard-copy letters or cold-call telephoning; it educates and informs clients; and it illustrates the local agent's level of expertise.

In early 2005, the NAR introduced a similar but separate e-newsletter, called a Market Conditions Report that concentrated more on national, regional and local property market information. An "umbrella" version was sent out to each local agent but then the completed local version would be automatically "filed" online so that potential buyers or sellers surfing the internet and seeking property information in a certain area would quickly have their attention drawn to this highly expert local agent's market report.

Case study 3: the US — how technology is used to train estate agents

Training of estate agents in the UK is still in its infancy. Although the NAEA has instructed a small number of courses for some years, these are negligible compared with many other countries and attract only relatively modest attendance. After all, with no compulsion to have minimum pre-entry requirements, and no licensing or other regulatory framework, many would-be estate agents think, "why bother?"

Even the NAEA appears unsure as to the value of qualifications. Its own website uses these words to describe what is needed to work as ...

... an estate agent: "Whilst academic qualifications are always helpful and applicable, personal qualities and abilities are the most important aspect. You also need to be literate and numerate."

... a lettings agent: "Academic qualifications are again helpful and useful, although communication skills are the most important. The ability to deal successfully with all sorts of people while handling someone else's valuable asset is vital. You will also need to be numerate and literate, and, preferably, computer literate."

... a commercial and business transfer centre: "The most important qualification needed to be successful is an ability to get on with a variety of people and a desire to help them fulfil their plans. Although not essential, academic qualifications show a technical ability and can help reassure clients of your ability."

... an auctioneer: "Academic qualifications are not essential. What is essential is an outgoing personality, the ability to communicate — which includes listening to everyone around one and excellent organisational skills."

This is some way from the US' approach where the main educational organisation for the industry is the NAR, the NAEA's equivalent body (albeit one which has a far greater influence in the industry and with law makers and other trade organisations).

Its courses cover: Association Management, Business Development, Business Planning, Buyer Representation, Commercial Real Estate, Diversity, Environmental Issues, Ethics, Fair Housing, Finance and Investing, Human Resources, Speaker Training, Overseas Property Purchasing, Land Brokerage, Leadership Training, Management, Marketing, Mergers and Acquisitions, Negotiating, Networking, New Home Sales, Personal and Professional Development, Professional Assistants, Property Management, Referrals, Relocation, Resorts, Risk Management, Second Homes and Site Selection.

The NAR also offers extensive online training. Classes are conducted over the internet and a separate telephone line is used for the pupil to call the instructor. For NAR members these classes are free and usually last about one hour.

The NAR uses "internet web conferencing" software that it has developed, called LearnLinc, to allow an instructor's computer to connect to the agent's computer and to run presentations on it. So when an agent joins a training class, software is loaded onto his computer to make it compatible with the instructor's machine.

This is not an elaborate process — new students are asked to "join" their class 10 minutes before its scheduled start time to allow the downloading of this software. Once the computer is connected, the agent calls a freephone number that will be included in a reminder e-mail that is provided to him or her prior to the start of the class.

Case study 4: Australia — the estate agent and the conveyancing solicitor come together in the "Real Estate Lawyer"

In Victoria, there is an attempt under way to merge the roles of conveyancing lawyers and estate agents — ostensibly along the lines of what happens in Scotland but with a strong new technology twist.

According to publicity from the firm, Real Estate Lawyer:

We're changing the real estate industry in Victoria. Now it is possible to sell your property through a Real Estate Lawyer instead of a real estate agent, and save thousands of dollars. Our flat fee of A$3300 includes professional photographer, signage, internet listing, conveyancing and legal work. In addition, all negotiations are handled by a fully qualified and experienced real estate lawyer (including price and contract terms and conditions). At all times, your most valuable asset is in safe hands.

The firm makes much of what it describes as a fundamental difference between the Lawyer Estate Agent and the conventional commission-based estate agent.

> The commission estate agent is not a true agent, as the role of the commission estate agent is simply to bring the buyer and the purchaser together. As an introduction agent, the commission estate agent acts for both parties at the same time. This leads to the problem of conflicting interests. The Lawyer Estate Agent, on the other hand, represents one person and one person only — the vendor.

Its standard sales package offered to clients include:

- preliminary legal advice on all aspects of pre-sale disclosure, sale procedure, contract matters, conveyancing and general legal matters relating to real estate
- an appointment with a local photographer to take high-quality digital pictures of the property going on sale
- the preparation of photographs and details for use on the Internet and a window display card, all checked by the client before use
- the erection of a For Sale sign and its removal after the sale
- "pointer" signs on nearby lampposts and community notice-boards, indicating the nearby home on sale
- hard-copy details to be given to prospective purchasers who view the property
- preparation and updating of a digital contract on the Lawyers Real Estate website that can be downloaded by prospective purchasers
- legal advice on the Sale of Land Act disclosure issues and advice on appropriate elements of the Section 32 Vendor's Statement (the two key legal elements of any sale contract in Australia)
- preparation of hard-copy contracts of sales
- after-hours contacts for clients or purchasers requiring information or legal advice regarding properties for sale
- handling of the sale by "lawyer estate agents", including negotiations on prices and contract conditions
- collecting and holding deposits, then releasing them in accordance with the seller's requirements
- lodging formal requests with mortgages, and arranging for the discharge of any existing mortgages or caveats covering the property

- all conveyancing work and legal issues associated with the sale, including delivery of the titles to the sale on completion day
- notification of the change of ownership to local bodies.

The company claims — as it would — that feedback suggests most purchasers want to be able to meet and talk with vendors directly. This may be done over the telephone initially before a viewing is arranged. "The vendor is the only person who can answer initial queries, and confirm an appointment time in the same phone call," it proclaims in its advertising material.

Much of the firm's claims are likely to be PR spin but the fact remains that around the world there are attempts to combine the different elements of the property transaction process into something approaching a one-stop shop. Will it happen in the UK? You bet — watch this space.

Case study 5: Australia — mobile telephone technology

Australia's Hutchison 3G telephone service has formed an alliance with real estate companies to launch what is called the first "location-based mobile data service" for property professionals who are onsite making a pitch to win an instruction. Called Loc3, it works like this:

- Agents enter their location into Loc3, which is operated through Motorola mobile telephones or a laptop fitted with a wireless 3 NetConnect Card (a simple card, similar to a mobile telephone Sim card).
- This allows the agent to download maps and aerial photos (supplied by two Australian data suppliers, Echo Solutions and Mapshed); he can also call up sales histories, current listings, comparable appraisals and market information while at the potential vendor's property. And, because this is wireless, it can be done even on a large estate or in a garden where there is no physical modem link. What it requires is a wireless mast facility somewhere nearby — currently, this covers 40% of Australia.
- After providing a valuation, a series of photographs or even limited video footage of the property can be shot using a compatible Motorola telephone and these can be e-mailed to the office for turning into formal draft details, which then can be sent

to a vendor's printer, or the agent's printer if he has one in the car. If older technology is used, the draft can be faxed to the vendor.

Hutchison is working with one of the big corporate residential agents in Australia, First National Real Estate (FNRE), to develop new software for rental property management so that agents can e-mail inspection reports directly from a rental property to the lettings office or the landlord.

A three-month pilot by FNRE showed that each sales office using the first version of the new technology produced extraordinary working benefits.

"The phones running Loc3 have saved us at least two hours a day by enabling agents to download, enter and send data at high speed in the field and generate reports on the spot. Now that properties can be viewed online only a few minutes after the valuation, vendors and buyers have been very impressed," according to the agency's sales principal, Pauline O'Neill.

Case study 6: Australia — licensing estate agents and technology qualifications

Every estate agent in Australia has to be licensed by a local board, normally consisting of magistrates, representatives of local councils and chambers of commerce. The board has the power to conduct legal and financial checks on the propriety of the applicants.

The process varies a little from state to state but the Northern Territory is typical. The legislation outlining the Territory's licensing system does not make an exciting read to the outside world but deserves selected quotation here. It speaks at length of what makes "a fit and proper person" to become an agent but then it sets out (in legal jargon, alas) what the board that supervises the issuing of licences looks for in a prospective agent:

A person, not being a company or firm, is eligible for the grant of a licence where the Board is satisfied that:
(a) he has attained the age of 18 years
(b) he is a fit and proper person
(c) he holds the prescribed educational qualifications for the class of licence which is the subject of the application or has other prescribed qualifications or experience

(d) by reason of his qualifications and experience he is competent to carry on business on his own account as a licensed agent; and

(e) he will, when licensed, be carrying on business as a licensed agent within the Territory.

The qualifications required are set out thus:

Where an application is made in accordance with this Part, the applicant is entitled to be registered as an agent's representative where he or she proves to the satisfaction of the Board that:

(a) he or she has attained the age of 18 years

(b) he or she is a fit and proper person

(c) he or she holds the prescribed educational qualifications; and

(d) he or she will be employed by, or in the service of, a licensed agent as an agent's representative within the Territory.

The Board may waive the prescribed educational qualifications where, in the opinion of the Board, the applicant:

(a) holds educational qualifications that the Board considers to be equivalent to the prescribed educational qualifications; and

(b) by reason of the applicant's experience, is competent to act as a registered agent's representative.

The experience referred to [above] may have been gained before or after the commencement of this Act and either within or outside the Territory.

The board has to sanction which qualifications are deemed appropriate.

Those currently required in the Northern Territory include extensive exam-based qualifications in handling internet technology, spreadsheets and word processing as well as "traditional" agency activities, such as valuations, property descriptions and fair trading. To ensure new agents have the right qualifications, the board can give (or refuse) its blessing to any local educational authority that puts forward an appropriate new course to count towards estate agents' qualifications:

The Board may ... approve a course of competency-based training referred to in the notice. In deciding whether to approve a course of competency-based training, the Board must consult with and consider the advice, if any, of:

(a) the Northern Territory Education and Training Authority; and

(b) an occupational association or body that represents the interests of the occupation to which the course relates.

(3) The Board:

(a) must approve a course of competency-based training if the Board is reasonably satisfied that an agent who satisfactorily completes the course will be competent to provide the services of an agent that are of the kind to which the course relates; and

(b) must not refuse to approve a course of competency-based training on the ground that a person who satisfactorily completes the course may not have attained the standard of best practice in the relevant industry.

Every assistant, whether administrative or to assist with viewings, or to sit in show homes of new residential developments being sold through an estate agent, must also be registered:

(1) A person other than a licensed agent shall not act as, or carry out any of the functions of, an agent's representative unless he or she is a registered agent's representative and he or she acts or carries out those functions for and on behalf of a licensed agent.

(2) A person other than a licensed agent shall not, unless he or she is a registered agent's representative, hold himself or herself out by any means as an agent's representative or as being in the employment of, or as acting for or on behalf of an agent as an agent's representative.

(3) A registered agent's representative shall not hold himself or herself out by any means as being in the employment of, or as acting for or on behalf of an agent, unless that agent is his or her employer, principal or partner.

Case study 7: Canada — a (fiercely protected) multi-listing system

As with the US and some mainland European countries, Canada has a national multi-listing service despite the vast size of the country.

But what differentiates Canada's MLS from every other system is that the Canadian Real Estate Association (CREA) — which is the nation's equivalent of the National Association of Estate Agents but which has a much higher proportion of the industry among its members and which exercises a substantial grip over the profession — does not allow individual estate agencies to advertise their properties on their own personal websites.

This policy is enforced vigorously. In 2004, the CREA took out an injunction against one individual realtor, the Sutton Group in Quebec, because it had copied some property details from the official CREA website and re-framed them under Sutton's logo on its personal website.

The philosophy is clear, with most individual agents supporting it, that the MLS is a more effective way of selling homes and is easier to use by the would-be buyer.

Tom Boseley, of Boseley Real Estate in Toronto, puts it this way: "Unless the CREA successfully stands up to any challenges to the system we run the risk of losing the MLS. The MLS system across this country has been proved to be an outstanding tool for the consumer. That's who the MLS was designed for — it's a system that was paid for by realtors, and is supported financially by realtors, for the betterment of the consumer."

Case study 8: The Netherlands — another multi-listing system

Estate agency in The Netherlands is strongly regulated by private organisations as well as by the government, with a strong emphasis on the use of technology.

Residential sales are based primarily on a restricted number of multi-listing services, one of which has a 65% share of the entire country's sales market. A string of private institutions, as well as forming the bulk of the rental and sales industry, also act as professional standards regulators for the industry. They set out codes of conduct, entrance requirements and lobbying arrangements with the government and other public sector institutions.

This is because in many areas of the country, most notably Amsterdam, virtually all properties for sale or rent are on the same multi-listing service. As with the US, a high proportion of purchasers in The Netherlands use buying agents — currently about 20% but believed to be rising rapidly.

The main estate agency professional body is the Nederlandse Vereniging van Makelaars og en Vastgoeddeskundigen (NVM), which certifies and registers its members. Only NVM members have direct access to the dominant multi-listing system that the parent body operates.

Members of the NVM and two other large bodies (VBO Makelaars and Landelijke Makelaars Vereniging) handle 80% of the housing transactions, all through multi-listing services.

For this reason, the OFT in the UK, in its 2004 survey, noted that there was more consumer body and government concern about potential monopolistic abuses in The Netherlands than in many other countries.

Interestingly, many NVM estate agents have formed links with mortgage brokers who are sometimes based in the same offices and are now registered by the same body as agents — perhaps the way that British estate agents may want to operate when the different elements of buying and selling property move closer together through the HIPs?

The *www.funda.nl* site makes the complete NVM multi-listing service content available to consumers but there are also several smaller competing sites, including one set up by The Netherlands' office of Microsoft, which links estate agents and mortgage brokers.

Now there is pressure to merge the NVM multi-listing contents with those rivals — in other words, to create a national service along the lines of the one in Canada (above) that means buyers and sellers have a one-stop shop in registering their property details. Perhaps as significant is the fact that many Netherlands estate agents see this breakthrough as the time when their relatively low-fee structure can be increased because of the improved service they offer the public.

Case study 9: The Netherlands — developers meet buyers via online technology

We are used to developers in the UK encouraging buyers to apply early and off plan to get a larger choice of fittings and perhaps even a slightly reduced price. But new technology is used by developers and their estate agents in The Netherlands for far more fundamental purposes.

AM Wonen is a major Dutch builder and on its website *www.amwonen.nl* one can find more than 40 different developments. Each development has its own micro website with information on the locality, the housing types offered, and public services in the neighbourhood.

So far so good — but now comes the content that is markedly different from what one would get in the UK. For example, there is usually a room planner on each micro site, operated by the user's mouse and created to assist in the layout of a room, including where developers should install sockets and shelving.

Later phases of most developments also involve the creation of a virtual community of off-plan buyers, in the form of an internet chatroom set up and moderated by AM Wonen, where buyers correspond with each other and the developer.

It helps AM Wonen to find out what the major issues are and marks a striking difference from UK developers who, rightly or wrongly, have a reputation for providing insufficient after-sales care (a factor addressed by Kate Barker in her review of new homes conducted in 2004).

Many estate agents also offer buyers a subscription to an encrypted service on the website *www.wenswonen.nl*. Called the "Woonplanner" or Living Planner, it was originally created by the Dutch volume builder Heijmans, allowing individuals to enter their room and furniture dimensions and then arrange their items as they wish before moving in.

There are practical advantages for the buyers — if their furniture is too large, they can dispose of some items in advance of moving without incurring removals costs. For the estate agent and developer, there is the goodwill of giving their clients access to a useful tool and enjoying an enhanced reputation for customer care.

Another subscription service offered to clients by estate agents is a demographic and so-called "lifestyle finder" on *www.prizm.nl*.

This divides Dutch postcodes into categories based on more than 500 lifestyle and demographic characteristics, including age, income, reading habits, hobbies, possessions and purchasing habits — if you have to move to a new area, make sure you find out about it first.

Case study 10: Denmark — yet another multi-listing system

The Danes have a high use of the internet and, unlike in the UK, much property advertising has deserted print newspapers and appears solely on websites.

This has been encouraged by the practice of estate agents charging sellers for the costs of advertising their homes, meaning that internet ads (for the sellers and then their agents) are significantly cheaper.

The importance of good websites with rapidly updated property information and high volumes of other data (mortgage information, details of local and regional facilities and accurate, objective market reports) has encouraged estate agents throughout Denmark to join franchises and affinity groups to achieve economies of scale and to maximise hit rates on sites. Unlike in Britain, many agents also use conventional television advertising to draw attention to their websites.

But, in 2003, as a result of pressure from the government, all the agents were allowed to continue their own internet websites but were also obliged to pool their property listings on to one master site owned by the main industry organisation.

As a result, the site *www.boligsiden.dk* usually has between 35,000 and 42,000 properties listed at any one time — the site is well used, with over 300,000 hits each month.

This is by far the best-used property site and agents that are not registered through the main professional body are not allowed to promote their homes through the site.

In return, percentage commission fees are much higher than in the UK.

Case study 11: Mainland Europe — here come the Americans

One of the US' most successful realtors, and therefore one of the world's largest estate agencies, is looking to expand into mainland Europe, with highly sophisticated technology at the centre of its approach.

Coldwell Banker started in property development in San Francisco in 1906 although did not have its first estate agency offices until 1925. It expanded slowly through the US Midwest over the following 45 years but in the 1970s it exploded, with large numbers of offices throughout Atlanta, Chicago and Washington DC.

In the 1990s it went international, linking with Corporacion de Bienes Raices Servicios Globales SA to take it into the Latin American countries of Costa Rica, Guatemala, El Salvador, Belize, Honduras, Panama and Nicaragua. Then, in Israel, it joined forces with existing agents Zerf & Nir Ltd to form Coldwell Banker-Israel. It is in Singapore and the Caribbean too.

It employs more than 106,000 negotiators and managers, has 300 offices throughout the world, sold over 11,000 properties worth $1 million or more in each of 2002, 2003 and 2004, and cumulatively sells more than one million homes each year. The business is valued at over $380 billion.

After all that success, it is now our turn to be put under the Coldwell Banker spotlight.

The firm's *modus operandi* is that it decides to enter a country's residential market, opens a number of offices but insists on training staff in the US to encourage the same renowned approach to improve customer care and high technology that are part and parcel of the "American home-selling experience".

In early 2004, Coldwell Banker opened offices in Spain, France and The Netherlands. In November 2004, it expanded into Poland. In all of those cases, the firm took its new staff many thousands of miles to the unlikely town of Buffalo in upstate New York to train them in technology supplied by a local firm that exists almost solely to supply Coldwell Banker with ever-improving software.

The firm, Algonquin Studios, has signed a 10-year deal with Coldwell Banker Europe to supply multi-lingual software (called CBE Advantage) so that every office across the continent will operate in the same way using the same techniques. By 2010, Coldwell Banker will have spent $20 million on training and software, with Algonquin recruiting an extra 125 staff to meet the demand. In total, CBE Advantage will be operating in 27 European nations, including the UK, by 2010.

Incidentally, in case you thought Coldwell Banker's expansion to Europe was a vivid example of how a big corporate can literally take over the world, you ought to know that Coldwell Banker is itself owned by a firm called Cendent — which also owns Century 21, the largest estate agency in the world, and also owns a more modest-sized US agency called Electronic Real Estate Association.

Can UK estate agents really believe they can survive this global onslaught with little or no technology, and little or no training, when compared with Coldwell Banker and the other giants vying for business in the UK?

Is it good enough to rest on the laurels of what the industry has achieved over the past 100 years of selling homes for people?

The answer is no. So what can we do about it?

The UK Should Not Fall Behind — What Is Available To Us?

There's no reason why a single estate agency office, excluding the property costs, couldn't be set up with state-of-the-art technology for about £8000, including both the hardware and software. If you then have a second office, the costs drop down significantly because the same software can be networked. Within a few months, they would get enough extra instructions and more property management clients than they would need to cover the cost. It's that simple — why don't people do it?

Those words come from someone with a clear vested interest — a software marketing executive whose firm works with estate agents. Although this statement makes it all seem simple, there is a kernel of truth in the fact that, despite the wealth of software and hardware now available, relatively few property professionals buy or lease them — and those who do, update them only rarely.

Different companies manufacture different software although, in essence, many perform similar functions. These tend to be split into various categories.

- *Property Portfolio Management* — that is: (i) the management of sales and the attendant paperwork concerned with instructions, offers and completions; (ii) the management of long-term or short-term lettings, again with the attendant paperwork, including time-specific letters reminding landlords about inspections or reminding tenants about service charges or late rent; and (iii) marketing, ranging from the basic display cards for branch windows to submitting pictures and copies to the local press.

Most Property Portfolio Management software also offers what is formally called Management Information (that is, data to help the running of a company and its staff whether that is one office and one agent, to a national chain with hundreds of offices and thousands of personnel).

- *Internet marketing tools* — this includes the creation, management and maintenance of an agency's own site, through to the uploading of material on to the major property portals, or even in some cases "swapping" properties with other agencies that are in the same affinity group.
- *Physical tools* — this is the real bread and butter of estate agency, but the new technology twist is that lasers and mobile electronic notebooks are replacing tape measures and pencil and paper.
- *New media* — this is the brave new world in which telephones and interactive television are used to promote properties or publicise the merits of one agency over another.

The rest of this chapter will give details of various types of each product. Inevitably, each example quoted below will be just one manufacturer's version of the product. They have not been chosen to offer particular endorsement, but simply because they are proven good examples of what is available on the open market.

In reality, there are dozens of rival products available in each category — many of the manufacturers are listed, with contact details, in Chapter 7 — and, ultimately, each individual estate agent must make a choice specific to his or her needs.

Property Portfolio Management and Management Information software

Sales management software

One of the leading management information systems available to the residential sales and lettings industry is Vebra's Estatecraft. It may seem remarkable that less sophisticated versions of this package have been on sale for almost 20 years but the current version is as good as it gets in terms of an easy-to-use system.

Using a series of templates based loosely around the ubiquitous Microsoft Word word processing package, Estatecraft allows an agent to carry out professional "logging" of business in the following ways:

- to complete a standard property card layout with configurable type fonts, logos, sizes and styles to suit individual firms and offices
- to track and record the entire sales process from initial valuation to final sale, including mailings, viewings, notes, verbal and written correspondence and fees
- to allow progress-chasing across branches to be undertaken centrally if required.

As a direct by-product, also it allows high-quality sales material to be created because the software allows:

- direct links with digital cameras, scanners and colour printers for rapid production of full-colour sales particulars
- the insertion of standard paragraphs about, for example, terms and conditions, directions, location maps, the estate agency logo and so on
- the creation of optional links to high-quality glossy brochure printing via a third-party print house — in other words, it can send property details on an e-mail pro forma, which will trigger an order for a particular type of brochure produced only by a specialist printer, perhaps in a different part of the country
- the creation of additional window cards or press advertisements about a property without the need for re-keying of details — this means it can automatically use the words typed in for hard-copy details to produce a window display card, or it can send those words by e-mail for a press advert
- the automatic incorporation of company or association logos, avoiding the need to use pre-printed letterheads
- these details to be automatically e-mailed to pre-set groups, such as other branch offices or subsets of clients
- the insertion of links to mapping software, allowing the automatic production of street maps, which can, in turn, be printed on property details.

For firms interested in long-term business management reporting, this software can also:

- record details of historical and current valuations to build a data-base of comparables, which can be accessed by any agent in the organisation
- allow powerful "matching" of variables to put would-be buyers

Detached double fronted family house built 1925 with spacious rooms refurbished and decorated to a high standard. The property retains much of its period character with wall panelling and original fireplaces. The rear garden is very large 203' x 60'3 (61.87m x 18.36m) and southwest facing. Eversley Road runs between Hermitage Road and Harold Road with the property facing a recreation ground with tennis courts. The nearest stations are Gipsy Hill and Norwood Junction. The property is located in Crystal Palace with its wide selection of shops and restaurants and is also well situated for schools in the private and state sector such as Dulwich College, DCPS.

This property management software from Vebra is typical of what is available. It can match property details to registered buyers, and simultaneously adapt the details entered at a branch to create the master database, window cards and press advertisements. The details and appropriate pictures can then be uploaded to the agency website and internet portals

Source: *Vebra*

and homes together quickly and accurately; this entails agents compiling pro formas about buyers and properties with similar contents in order to make accurate matches — the more details a company has about homes and people, the closer the matches

- allow searches to be made on a variety of criteria; for example, the client database can be interrogated to find out how many buyers want properties within a certain price range, or how many properties of a certain type and size are on sale at any one time, or how many properties of a certain type or size have been on sale historically through the office in question

- measure customer service through searches of data on the number of letters sent to buyers, or calls made to vendors, or numbers of viewings arranged, etc.

For those agencies adopting sophisticated electronic marketing campaigns, the software can help by:

Software for lettings agents, like this from Websky's Expert Landlord product range, matches details of properties to would-be renters and helps with practical management — a "scheduler" component will automatically generate reminders to landlords and tenants about safety checks or routine inspections, for example.
Source: *Websky*

- allowing property details to be uploaded automatically to a website
- then passing details to major portals with which an agent is linked (for example, Assertahome or Primelocation)
- e-mailing details and photographs of properties to clients, or sending draft details for checking by vendors
- creating mobile telephone texts to vendors, negotiators and would-be buyers.

Long-term rentals management software

Websky, another software provider for the property industry, provides a lettings program called Expert Landlord, which is one of the leading packages to make the handling of bulk lettings easy, reliable and, ultimately, more profitable.

It provides a series of wizards (the computer industry term for onscreen software guides and templates to help inexperienced users follow basic instructions) to demonstrate how to install and use the program.

Expert Landlord can:

- store details of a landlord's portfolio, no matter how large, complete with details of size, capital cost, maintenance issues, annual yield and so on
- hold stocks of compliance letters, notices and agreements supplied by a lettings law firm and to be issued to tenants and landlords in different circumstances — the dispatch of these can be triggered by date to avoid agents forgetting that periods of notice are expiring, for example
- set automatic reminders about gas checks and safety inspections and link them to the electronic diaries of individual agents or landlords
- hold and print photos for lettings details, window cards or press advertisements
- create property lists compiling all units available for renting, irrespective of the number of owners of those properties
- store purchase price, current value and mortgage details to help calculate portfolio values — these can be automatically updated to take account of RPI, local property inflation, market changes, etc
- create and monitor key-control systems to assist with security
- use a cashbook function to control rent payment, expenditure and outgoings
- generate figures for tax returns and integrate with online insurance and reference-checking agencies.

These facilities can be modified to handle activities related to room-renting as well as separate properties, and can apply to tenant-sharers as well as traditional individual renting.

In addition, there are eviction options, showing the legal alternatives open to individual landlords or agents, again using information provided by an associated legal adviser.

Short-term holiday lets software

CFP Software has a version of its long-term lettings management software modified for holiday lets. This is a growing part of the estate agent industry now that there are over one million second homes in the UK, of which about 45% are let out for parts or all of each year to weekly or fortnightly bookings.

The CFP product allows a member of the public to log on to a holiday let agency's or owner's website and nominate a chosen week, fortnight or other preferred holiday period. This request is then passed through to the lettings agency for confirmation.

In theory, the person could simply make the request and present his or her credit card details and the booking could be confirmed automatically, but the software firm claims the vast majority of agents prefer this manual dog-leg route to allow them to personally vet the financial bona fides of the client who is booking.

When the lettings agent is in a position to confirm, the chosen dates are manually entered into the "scheduler", an electronic diary that lies at the heart of this software application. A quick drag of the mouse makes the reservation and calculates the deposit required and the full holiday cost (assuming that the scheduler has been updated with the differential rates for different seasons plus any optional extra charges). These costs can then be sent, electronically of course, to the client making the booking.

The software also has a complete double entry client accounting system, an organiser to help users keep track of day-to-day administration, including void periods, and a tick-off bank reconciliation facility. The software can generate mobile telephone text messages (for example, to be sent to the person confirming a reservation, or to a landlord saying his or her property has been booked); it can also produce window cards for those estate agents publicising holiday lets.

As with much long-term letting software, this package can offer automatic cheque printing; a BACS logger that records electronic payments to landlords and contractors; the collected archives of owner, property and enquirer information; and fully automatic letter printing — this includes batches of notification letters to those making bookings, or letters to contractors.

The scheduler can track periodic events, such as the need for health and safety checks, insurance renewals, cleaning, gardening or redecoration, and can notify the landlords, tenants or contractors by letter, e-mail of even mobile telephone text as appropriate.

CFP recommends that training begins at the agent's office, loading real details of owners, properties and bookings on to the agent's computers. A total of three sessions, each of three hours, is suggested;

there is also a telephone hotline and updates program to help with queries from new users and the installation of software upgrades.

CFP says its biggest client has a 20-screen operation, each used daily during peak booking periods to handle a portfolio of over 500 lettings properties; its smallest client is an individual landlord who has 12 properties that he manages himself. The latter cost just £3500 to set up with the software, training and first year's support program.

Housing Association software

This is not as irrelevant to the private market estate agent as it may seem at first sight.

An increasing number of Registered Social Landlords (RSLs), to use the official term for housing associations, are engaged in forms of shared ownership whereby a proportion of a property is owned by the association and the rest is bought on the open market by a former tenant or even a general buyer. The purchaser then pays a conventional monthly mortgage for his or her "share" of the property and a monthly rental payment to the association for the remaining share.

When it comes to selling the property, any appreciation (or depreciation if the market has fallen) will be shared proportionately between the private owner and the RSL. Look out for this concept spreading to the mainstream housing market, as several private developers recognise that many first-time buyers cannot afford a "full" property but could afford a share of one.

Therefore, specialist software for RSLs, such as that developed by Odyssey Interactive, may give useful lessons to the private sector in the near future too.

Odyssey's Interact Housing package allows housing associations to develop an intranet (a computer industry term for an internet service accessible only to a few restricted users).

The intranet is tailored to each organisation's corporate branding and design, and is based on a "core system", consisting of a searchable staff directory listing all employees from one central location. It boasts easy-to-use content management that non-IT literate individuals can operate, including fully configurable menus — another piece of jargon, meaning that each individual gets their own page on the screen when they log on. There is also a high level of security, restricting other individuals, or departments in the organisation, from accessing certain information (so not everyone can see buyers' details or tenants' financial references, for example).

RSLs can add or take away components from this core system with an unlimited number of additional modules, including a "helpdesk module" that enables organisations to keep on top of tenant enquiries, and an "interactive forms management module" that allows users to create and submit forms for new properties or building repairs, and so on.

There are clear potential uses for the traditional private rented sector here too, according to Odyssey Interactive's Nigel Danson. "Because all information is stored from one central location, every employee has easy and direct access. Because everyone has access, the costs of printing and distributing information are greatly reduced."

The software is already used by Chester and District Housing Trust, Moss Acre Housing and Gallions Housing Association — the latter adapting it to help manage and develop over 5000 homes in the Thamesmead area of London plus the maintenance and development of lakes, waterways and parkland in the area.

Gallions says it wanted to improve the sharing of information between staff and reduce the cost of processes, such as expenses and room and resource bookings.

The Interact package was chosen because it enabled the organisation to invest in only the essential functions at the outset, with the option to add new functions when required and within a restricted budget. Crucially, Interact could also be implemented within a short timescale of just two weeks.

Gallions installed an intranet that contains a small number of business applications to encourage regular usage by staff. This includes sharing information within teams relating to regular procedures, such as ordering, as well as speeding up processes, such as handling expenses and payments or managing room bookings for residents' meetings or agreeing diary dates with builders and contractors coming on site. The package was tailored to give specific details to each department within the association, so staff did not have to scroll down a large volume of information to see the part relevant to that individual or team.

Gallions now claims that internal communication has been improved, and that all the content can be updated by non-IT staff, allowing the IT team to concentrate on technical matters.

Postcode look-up software

There is one good reason for using this type of software — it saves some 80% of key strokes, which may be of little interest to some estate agents but may keep their administrative staff from tedium, repetitive strain injury and even discourage them from lodging the odd compensation claim.

Thesaurus Technology produces a postcode look-up software package that involves a user simply keying in the postcode, clicking on "Lookup" and then watching the address details be displayed with the cursor placed at the start of the first line for them to key in the house name or number.

If the address has unique use of its full postcode (in the event of it being a large organisation or a grand country house, perhaps) the owner's name is also shown. This not only reduces keystrokes, but also saves hours of wasted administrative time, limits spelling mistakes, creates reliable address lists and gives a professional image for a business.

Of course, it is only as good as the information put in, so a new directory is required in an area where there are new properties being built. But these are readily and cheaply available.

Telephone call logging software

Talk to James Barnes of Newfound Properties International (NPI) if you have any doubt as to the merit of this type of software. When he resigned as head of international property at Hamptons International to set up the London branch of NPI, he had to consider what technology to use.

His business needs were unlike most agents. He dealt only in medium to high-priced overseas properties, usually in exotic, far-flung locations and different time zones, such as Antigua and Newfoundland.

"We use a unique telephone number for each press or internet advertisement we place. All the calls actually come into our same telephones in our office and we answer as usual but the dialled number is logged so we can check on the effectiveness of each individual advert. When a full page in the *Financial Times* can cost as much as £14,000, even with a discount, the importance of such monitoring is crucial," he says. His system cost about £6000 but he calculates that the targeted advertising it allows him to do saved him that amount of money "in about a month".

Systems such as Oak's Advance Classic can analyse the time taken to answer calls at switchboard, department or extension level. The same firm's Voice Pro system is one of those sometimes-annoying automated voice systems, but it kicks in as and when you want it to. This at least ensures a client always gets his telephone call answered — and it can be programmed to put people through to departments, give them menus, offer different verbal options to known callers with identifiable numbers, and it can even automatically change the message according to the day of the week and the time of the day.

The TIM Professional package is even more extraordinary, logging calls as they come in and producing hard copy or onscreen management information reports that would assist any multi-branch estate agency in particular.

It can produce billing reports, allowing you to produce a specific outgoing call cost bill for a client if you ever need to itemise expenditure on the sale of any one particular property. Also, it can give individual extension reports, which records the cost of every call from every extension or group of extensions if appropriate — again, this can be used to impress clients on the effort put in to a sale, as well as being used for managing staff costs if an employee is spending too long on the telephone.

The same package produces time reports that break each business day into half-hour periods, showing the maximum and average number of telephone lines in use during any particular half hour. It also has "call geography" features to give detailed information about where an office, extension or individuals makes calls to — local, national, international, mobiles and so on.

All of this information can be broken down further to include "top calls", which may be those numbers most frequently dialled, the most expensive or the longest.

All of the above is for calls emanating from an office but TIM also offers incoming call analysis, again, broken down by time of day, the time taken to answer them as well as call duration and caller ID.

Internet marketing tools

You may think there is little more that can be done to improve estate agents' use of the internet. UK agents are genuinely prolific users of the "net", but many sites are very traditional. They merely display properties and usually give fewer details than are available if a visitor walks into a branch of the same agency and asks for hard-copy printouts of the details.

There is logic to this of course. Getting the client in to an office may allow some additional marketing opportunities of those properties in other locations or in higher price ranges, which cannot logically be included on a remote internet search.

Data from the US shows that this may be a false economy. Surveys conducted for the NAR as far back as 2000 show that, because of the internet, people buy properties in a very different way.

Internet users tend to physically view fewer properties — 8 on average compared with 15 if they do not use the internet at all. Internet users are also much more informed about the market and availability, because, despite having the time-saving of short-listing via the internet, they spend three times more studying properties than non-internet users spend.

Therefore, this section looks at specific tools that can be used to enhance the "online experience" for clients.

Keywords

Keywords are the 21st century equivalent of the *Yellow Pages* — internet users feed them in to search engines to get the information they require. But the responses picked up by the search engines (that is, Google, Ask Jeeves, Yahoo! and so on) depend on which keywords the site owners (that is, you and your rival agents) have nominated to the people who create the websites.

Look at it this way. If I were moving house from London to Edinburgh, I might call up an internet search engine such as Google (which is easily the world's most popular engine) and type in the words: "estate", "agents" and "edinburgh".

The estate agents' websites that are listed, and the order in which they appear to me in the results pages returned to my PC by Google, will depend on the keywords fed in by the individual websites and then subsequently scoured by Google. So, even though the words "estate', "agents" and "edinburgh" seem basic enough, if a large Edinburgh estate agency with a glamorous site only nominated "scotland" "houses" and "flats" as its keywords, the site may not be listed on Google's results for my search.

Before there is any further discussion about keywords, there is an obvious point to be made here. The more professional and experienced the individual or firm that creates and manages an estate agent's website, the more likely it will be that the site will have its keywords tailored and then refreshed to ensure it is picked up by the search engines.

So a good "webmaster", to use the patois of the internet industry, will perhaps ensure that the keywords "edinburgh", "estate", "agent", "agency", "homes", "scotland", "apartments", "flats", "tenements", "houses", "semi", "semi-", "detached", "semi-detached", "mansion" and many others are inserted into the code to cover all options. Some clever webmasters even insert misspelt words as keywords — for example, one estate agency in Middlesbrough shrewdly uses "Middlesborough" as a keyword.

There is little benefit in getting too technical in this section, but keywords that are scoured by search engines tend to be in two categories. They have, in internet jargon, two types of "meta tags'.

One is a description meta tag that is usually under 20 words in length and is normally in prose English, and describes the business of the website — for example, "estate agent" or "selling homes". Some search engines operating in the English language look at this tag first and will prioritise results based on its contents.

The second meta tag is a simple list of up to about 40 keywords that do not need to be in prose style and can look to the casual observer like a random list (this would be where we would put those keywords from our example — "edinburgh", "estate", "agent", "agency", "homes", "scotland", "apartments", "flats", "tenements", "houses", "semi", "semi-", "detached", "semi-detached", etc). Some search engines look at this particular tag first and will prioritise results accordingly.

The reason that the two types of tag are both required is that there are different types of search engine used by us all when we use the internet.

One is a "catalogue-style" search engine, such as Yahoo!, where webmasters can choose the section to which they submit a website — so they can submit an estate agent website to, say, "estate agent" and "selling homes" in the case of our example.

The second type of engine is a "crawler site", such as Google. These engines are more sophisticated and when a user feeds in a word they send out a "spider" to crawl through sites in a much more indiscriminate way — therefore, the individual non-prose keywords are especially critical.

Most estate agents with little further interest in the technology behind websites need know no more than this. They merely have to ensure that their webmaster is experienced and good enough to produce tags that mean their websites comes up on the widest range of searches conducted by the largest number of search engines.

Now let us stay with the internet but move on to something less technical.

Enhanced search facilities on websites

If you think all property website search facilities are the same (because they all seem to want users to feed in location, price, number of bedrooms and the like), then think again.

From early 2005, Vebra has been offering "descriptive searches", allowing the public to put in flowing text, thanks to new software developed initially by the people behind Google. Vebra's development director, Simon Mackin, says that once a user is on a property website it has been a tradition to merely use fixed parameters, such as price, location, property type or number of rooms to generate results. "But now consumers can enter descriptive searches, for example "three-bed semi in Camberwell" or for more unusual features, styles or specific locations like "barn conversion with conservatory and views of the Cotswolds".

The new search system operates in the same easy-to-use way found on Google itself, with quick results too. "Offering parametric as well as keyword and description searches will enhance the way buyers use agents' websites, giving wider search capabilities," says Mackin.

Or, in English, vendors' properties will receive more exposure to the right type of buyer by extending the vocabulary that the internet user can apply to a search.

Internet advertising

This is not new but it is rising in influence. Figures from the Interactive Advertising Bureau show that online advertising in 2003/04 was 80% higher than in 2002/03 and that, when measured by cost, it accounted for 2.5% of total UK advertising — twice as much as spent on cinema advertising and over two-thirds of the sum spent on radio advertising.

One developer, called Modern City Living, has adopted an unusual technique for placing highly cost-effective "pay per click", or PPC, advertising for apartments that it is selling in central London.

PPC works like this: background software analyses keywords fed into certain search engines by users. In this example, Modern City Living determined that its keywords would be "property", "london", "city", "living", "apartment", "modern apartment" and many others.

For a fee, the search engine (in this example Google) agrees to put Modern City Living banner advertisements on its results pages that appear in front of the internet user who typed in the keywords.

The advantage for Modern City Living is that, as an advertiser, it only pays the agreed sum if the user then clicks on the banner advertisement and subsequently goes into the developer's own website. The counting of such "click throughs" is done by an agreed automatic method, which makes it far more cost-effective and accountable than regular internet advertising.

There are literally hundreds of internet firms offering to handle PPC. The world's largest is Overture, used by Modern City Living. It allows a client to barter the fee for each click through, so if the amount of traffic that comes to the client's website is lower than expected, a reduction can be negotiated with Overture.

Modern City Living saw a three-fold increase in the number of "hits" on its own website (the computer term for individual visits) from 8900 in January 2004 to 24,651 in June 2004. The average number of pages looked at per visit also rose, from four to eight. As a result of the success of the advertising, Modern City Living extensively upgraded its site to include downloadable brochures, computer graphic images of its developments, floor plans and virtual tours.

Another industry player, Newfound Property International (NPI), is placing a different form of internet advertising — by paying for banner advertisements on National Health Service (NHS) "intranets", which are the internal databases and websites used by individual NHS trust staff and patients in different parts of the UK.

NPI calculates that many of its users will be better-paid NHS staff, such as doctors and consultants, and, in some cases, more technology-savvy patients. The advertisements will emphasise the potential for international properties to complement or replace pensions... perhaps just what you want to think of when facing that check-up or operation.

Creating virtual tours on your website

e-house was one of the first companies in Britain to create virtual tours (VTs), the moving image displays of room interiors and sometimes exteriors that are available for putting on websites.

e-house's product differs from many of the current suppliers of this service by using its own professional photographers (it says this is to avoid relying on the sometimes-indifferent quality of an agent's own photographic skills). The e-house photographer takes between 20 and

By definition virtual tours are difficult to demonstrate in a book — they are moving images. But they are now easy to create. This example, produced by e-house, gives potential buyers a more interesting idea of how the outside space looks compared to a traditional still picture on a website.

Source: *e-house*

This internal tour gives a real feeling of space. Virtual tours are good on any property but those agents that use them say they come into their own when showing contemporary design interiors.

Source: *e-house*

30 pictures of each room (some perhaps with a fish eye lens, others with a conventional lens). These are then technically stitched together to allow a user to have a complete panorama with similar colour, textures and lighting throughout.

e-house claims that it can ensure the virtual tour is ready just one working day after the pictures are taken, irrespective of whether the property is "in Scotland to Cornwall, East Anglia and even West Wales," according to a company spokesman.

The system is rapid and goes some way to addressing the major complaint about VTs — that is, that they take a long time to download even to a computer accessing the internet through broadband, and are therefore unappealing to the casual user.

e-house claims to have developed a system called background loading, which means that, instead of a user waiting for the entire file of images to download, the first few images load quickly and can be watched by the user as subsequent images are still loading.

A rival estate agency support specialist, Estate Agency Software Solutions, offers what it calls Easpanoramic. Central to this is a simple 360-degree lens that, the firm says, allows an agent to create a virtual tour from a single picture.

Once uploaded to either of two software packages marketed by the firm, the virtual tour can be live on the internet with 24 working hours. Up to five tours can been added per property (to be used on the five most desireable rooms, presumably, or four rooms and an external shot) at a cost of £7.50 per property.

The lens is compatible with 12 different, mainstream camera models from Nikon, Sony, Fuji and Canon.

Whichever package you select, there are novel ways of using the technology. One estate agents, Stanifords, even puts virtual tours of its three offices on its website *www.stanifords.com*.

Putting floor plans on your website (and on hard-copy property details too)

Floor plans are becoming increasingly popular with agents and buyers, and a few agents even give the price of older second-hand homes in pounds per square foot. Although this is still only occasionally the case, it indicates this way of measuring size and cost is no longer considered the province of new-build properties only.

Acquaint CRM, a software supplier, offers a remote service that relies on the basic measurements being taken by the estate agent on the spot; measurements are then passed to the company, which makes electronic floor plans and e-mails them back to the agent for checking. The process works like this:

- The estate agent makes a sketch of each floor of the property, including dimensions in metric or imperial units, with each room identified (dining room, kitchen, etc), and clear indications of the locations of windows and doors (with notes of whether they open in, out, to the left or right) plus locations of fireplaces, bathroom fixtures, kitchen units and appliances.
- These measurements are faxed, posted or e-mailed with the sketch to Acquaint CRM.
- After a maximum of two working days, a draft of the plan will be returned by the company by e-mail, fax or post for correction or amendment by the agent. The plan will include all details, both metric and imperial measurements (even if only one has been supplied) and can be produced in the client agent's corporate identify, including fonts, colours and watermarks as appropriate.
- Plans are drawn to a 1:100 scale and automatically cropped to the ordered document or internet display size.
- After the agent's amendments to the draft have been integrated, the final version will be sent by Acquaint CRM, again by e-mail, fax or post. A version will be sent for inclusion in hard-copy printed material to go to clients, and an alternative size will be selected and e-mailed for inclusion on the agent's website and portals.

There is substantial evidence that adding a floor plan layout is not merely a cosmetic exercise to improve the appearance of a dull hard-copy brochure, or to add spurious authority to property details. Former estate agent Bob North, now a management consultant who has also worked at the property website *www.assertahome.com*, wrote in *Estate Agency News* in summer 2004:

> Niche [a software provider] is now providing a nationwide service, with a recent office opening in the North giving coverage up to the Scottish borders. Peter Burnham, their director of digital imaging services, is delighted at the increased number of inquiries it is getting from new areas.
>
> Recent agents starting to use the service have included Anderton Bosonnet in Lancashire's Ribble Valley and Palmer Snell in Dorset. Peter backs up the view that buyers want actions not words and is a supporter of [virtual] room tours as well as photos to give a lasting impression. His view

ground floor   *first floor*

GROUND FLOOR

Hardwood door to ENTRANCE HALL: Understairs storage cupboard, radiator, telephone point, decorative dado rail, stairs to first floor landing.

LOUNGE 5.03m x 3.25m (16'6" x 10'8"): Telephone point, decorative dado rail, two wall lights, coving to textured ceiling, double glazed box window to front, open fireplace with marble surround and wooden mantle over, two radiators, double door to:

DINING ROOM 3.02m x 2.74m (9'11" x 9'): Decorative dado rail, coving to textured ceiling, radiator, double glazed patio door to garden, door to:

KITCHEN 3.02m x 3.40m (9'11" x 11'2"): Fitted with a matching range of base and eye level units with worktop space over and underlighting. 1½ bowl stainless steel sink with single drainer and mixer tap with tiled splashbacks, plumbing for dishwasher, space for fridge, built-in electric fan-assisted double oven, built-in four ring gas hob with extractor hood over, laminate flooring, telephone point, decorative dado rail, coving to textured ceiling, double glazed window to rear, door to:

UTILITY ROOM 1.78m x 1.52m (5'10" x 5'): Fitted with a range of base units with worktop space over, 1½ bowl stainless steel sink with single drainer and mixer tap, plumbing for automatic washing machine, space for fridge-freezer, laminate flooring, dado rail, coving to textured ceiling, double glazed window and door to rear, door to:

CLOAKROOM: Fitted with two piece suite comprising wash hand basin and low-level WC, dado rail, coving to textured ceiling, radiator.

INTEGRAL GARAGE 5.05m x 2.44m (16'7" x 8'): Shelving, wall mounted gas boiler serving heating system and domestic hot water with heating timer control.

Floor plans were once only the preserve of upmarket new homes. But including them on the detail of any property offers a real service to buyers and shows an agent is going further to demonstrate the qualities of a home on sale. This example is typical of those prepared by a third-party company using measurements supplied by an agent.
Source: *Mobile Agent*

on fees is that it is a good tool to help to justify a full fee and can make all the difference in getting the instruction.

This view is also shared by an agent from south of Manchester, who wants to remain anonymous — he doesn't want his local opposition to realise how well he is doing by offering the full service. "Not only am I getting more instructions but my fee percentage has gone up by 17%," he said.

Using the highest-quality digital photography

There's nothing to taking pictures with a digital camera, eh?

Tempting as it is for many low-budget estate agents to apparently save money by learning how to use the latest generation of digital cameras, the results can be amateur. Put bluntly, the ease with which pictures can now be taken, transferred to different formats or printed out does not mean that we have all learned more fundamental activities, such as how to focus the lens to begin with, how to frame a shot properly, or what height to stand at to optimise the property's appearance.

Therefore, numerous specialist firms offer expert property photography, with shots taken in ultra-high resolution.

For example, e-house e-mails all photographs of a property to the sales agent in medium resolution in order to keep the size of e-mails manageable. But agents can then obtain high-resolution hard copies at any time, 24 hours a day, via an encrypted online service. These higher-resolution versions can be used for brochures, advertisements or even to accompany editorial in local or national newspapers or magazines.

This latter point should not be underestimated. Many of the large estate agencies employ highly professional and well-resourced public relations firms to try to win "space" for properties in the burgeoning numbers of property supplements in weekend national newspapers and lifestyle magazines. These publications often choose to feature a property solely on its appearance and the quality of pictures supplied by the agent.

For example *What House?*, which is possibly the leading and certainly the longest-standing magazine for would-be home buyers in the UK, insists that any photograph meets stringent standards before even being considered for use.

The magazine's "dope sheet" for contributors is typical of what modern publications want. It specifies:

* All pictures supplied must be able to be reproduced at A5 size and at a resolution of at least 300dpi.

- 35mm and medium-format transparencies/slides are the ideal format.
- If using a digital camera, the size setting should be no less than 3.2 megapixels (that is, the maximum available size with a resolution of 300dpi), allowing the pictures to be reproduced at A4 size.
- No pictures can be accepted at less than 300dpi. This means that pictures taken from websites or digital cameras on a setting less than 3.2 megapixels that reproduce pictures of just 72dpi are not acceptable.
- If supplying photographic prints they must be developed at no less than 8 inches × 6 inches (155mm x 205mm) or, ideally, bigger. Naturally, the main subject of the photograph must be clear and sharp.
- If scanning photographic prints for e-mail, they must be done so at A5 size or the high resolution setting on your scanner. They should be saved as j-peg or tiff files. Ideally, you should supply the original pictures so that we can scan them ourselves.
- Ensure that the subject of the picture is well framed and lit. Readers do not want to see partial pictures of houses in cloudy places if they want to buy a dream home in the sun!
- Ensure all pictures supplied are clearly labelled with the following: type or model of house (if applicable); cost of house; details of house (location, number of beds, baths, special features eg, swimming pool etc); your company contact details (full telephone number, including country code, e-mail, web address).

The point of reproducing this crib sheet in full is not to encourage estate agents to understand how digital photography works, nor to become semi-professional snappers. It is to show that high-quality pictures can pay real benefits by at least giving properties a chance of receiving editorial coverage. A feature in a newspaper or magazine is effectively free of charge and many believe it to be more influential in terms of getting enquiries than paid advertisements in the same publications.

Sometimes agents can buy photographic services from facilities firms that also handle other services.

For example, a company called Big Property Marketing has a network of professional photographers familiar with the Property Misdescriptions Act. For a fee of £100–£300, they photograph a property and create a floor plan and e-mail the images to the agent for insertion into the printed details.

Incidentally, hiring a professional firm to produce high-quality digital pictures has one other benefit. Modern touch-up facilities ensure that all external shots have blue sky, no matter how bad the weather was when the picture was taken.

Advanced internet techniques, such as Micro Websites, especially for new homes

Residential developers often require similar packages of technology to those sought by the more ambitious estate agents, but with subtle differences and enhancements.

e-house offers a package of electronically delivered services, including professional photography of properties, virtual tours, brochure design and, most innovatively, the registration and creation of "micro websites".

Within the property industry, a micro website is one that is dedicated solely to one new development and it usually carries a domain name (that is, the exact internet address that identifies the site) that is the same or part of the development's title.

For example, St James' Homes has its own website (*www. stjameshomes.co.uk*) and within it there are sections for each of its current developments.

But Wycombe Square near Notting Hill in west London, one of St James' largest and most controversial developments (which has townhouses with prices exceeding £10 million and which caused a furore when it went ahead to the consternation of well-heeled residents living nearby), has its own micro website. This is found at *www.wycombesquare.com* and is dedicated solely to that development — its address is also featured on its marketing and advertising, allowing people to "enter" the micro site without first going via the St James "macro" site.

This concept offers a number of advantages. First, in terms of public relations and image, it can look highly prestigious to a would-be buyer to see that a development has what is perceived as its "own" website.

Second, a separate identity can give more space to details and images of the development than may be possible on a developer's generic website, which has to carry information on possibly dozens of other developments.

Third, automatic counting of visitors to a unique micro website will be more accurate than trying to guess how many visitors to a generic site looked at those pages given to a specific development. Information of this kind (called "site traffic" in the industry) is often used to gauge interest in a particular development, or even to satisfy anxious shareholders who may be concerned at the level of publicity a development is generating.

21st century offices and tools

Everything we have looked at so far has been software or software-related. But simple physical tools can also be used to make an estate agent's work easier and more accurate too.

Laser measurers

Swiss precision tools firm Leica manufactures a range of hand-held laser distance meters in its "Disto" range; most use a visible red laser for measuring up to 200m, or some 640 feet, with an accuracy of plus or minus 3mm. Some have luminous displays to show measurements, while others have graphic displays.

Some have calculation functions to immediately calculate the floor space of a room or a total property.

US engineering firm Sonin makes a Multi-measure 60 Pro-Plus, a versatile unit that, with one click, operates a laser measurer operating at short distances to a maximum of 18m. It has what it calls a "smart beam cone", which will measure down narrowing crevices or extremely awkward shaped rooms — ideal for establishing the maximum distances in unusual properties.

PDAs, wireless internet and on-screen agency activity

Personal Digital Assistants (PDAs) are those hand-held "mini PCs" that are available in high street electrical goods stores or via online ordering from the internet; they are fast becoming reliable substitutes for paper notebooks and old-fashioned diaries, and can even act as mini PCs because they can carry internet connections as well as most common word processing and office software packages, such as Word and Excel.

One major benefit is that most standard PDA software is compatible with most office computer software, so the PDA contents should be downloaded to the office computer regularly to avoid problems if the PDA itself is lost.

PDA screens are small so the period of time someone can use them is likely to be limited by simple tiredness, but they are fantastically light and small — some are now even being combined with mobile telephones (see Smart Phones below). PDAs cost from around £150–£400, depending on features and sophistication.

The second major benefit is that most PDAs are capable of "wireless internet connection", which will revolutionise working practices for many estate agents.

Most PDAs have built-in transmitters or a wireless modem (literally a modem with no wire and therefore not requiring plugging-in to any physical socket). In many areas, signals from the transmitter or modem will be able to connect with a wireless facility so that internet and e-mail links can be established.

Wireless connectivity does not work everywhere. The more heavily populated areas usually have it but connectivity is poor in some rural areas — a small onscreen message on the PDA will tell you whether a signal works.

At least one company has developed special estate agency software for PDAs.

Called "Mobile Agent", this comes from the Northampton-based Prototec, a firm majoring in floor plan drawing. Using the stylus that comes with a PDA, the agent can create the shape of each room in a property by using pre-installed templates that can be easily modified to depict even the most eccentric room shape. Then, the user types in the dimensions (achieved through a conventional tape measure or laser measurer) and slots in doors, windows, fireplaces, radiators and every other standard floor plan element by double clicking on icons at the top of the PDA or tablet screen.

The agent can also feed in other information, such as room titles, details of furniture and accompanying descriptions, and have the floor plan complete while still at the property. Then he can slot the PDA into a stand at the branch office or in his or her home, in order to download all of this data to a conventional PC. Using complementary Prototec software, this data is converted into PDF format for e-mailing to a client or interested would-be buyer, or it can be dropped into an existing Word file for the creation of regular hard-copy details on a personal computer.

The magic of this is that it does away with the need for painful manual creation of floor plans by hand, often done on graph paper

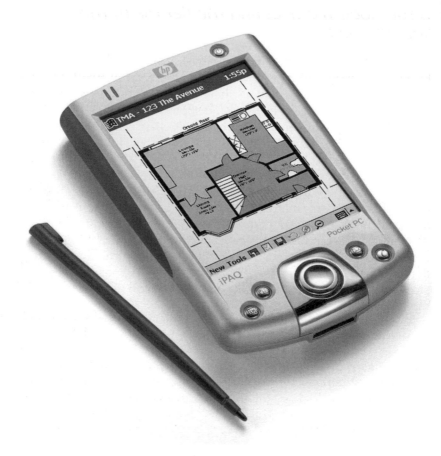

Now floorplans can be created by an agent during a visit, using a hand-held Personal Digital Assistant. Eventually, the data is downloaded to a conventional PC to create the hard-copy plans, or to be transferred to a website. But how impressed would a client be to see an agent able to use this technology in their home?

Source: *Mobile Agent*

and then faxed or scanned and e-mailed away to a floor plan creation company.

And the cost is a snip at £399.99 plus VAT for the basic software.

Ultra-modern offices and the demise of the window card

Imagine the typical estate agency office — large windows littered with property photographs and details, the gaps between them allowing you a glimpse of two or three desks, each probably with a telephone, personal computer and a small mountain of paper. Oh, and a negotiator sitting behind each desk too.

By contrast, imagine an upmarket coffee shop in Kensington — a glass table, leather seats, perhaps a magazine or two, with people enjoying a latte.

Now combine the two, and you have the most attractive estate agency ever. That is what a few agents are beginning to do in central London.

Foxtons — an agency with a controversial reputation because of its overtly ruthless, commission-led approach to selling homes — claims to have led this new approach with examples like its office in Ealing, which was opened in late-2004 and which is to be the model for a roll-out of similar refurbishments in other offices taking place until the end of 2007.

This includes the coffee shop-style frontage to each office, complete with coffee; just behind there is a "dealing room" of negotiators, each working at a "grid" of tables with equipment installed by Lane Business Systems. The equipment includes plasma screens showing up-to-date property details through a direct link to the Foxtons website; this will be the main display tool for use by potential buyers who visit the offices. For each property they will be able to see high-resolution colour photos, full property details, floor plans and 360-degree virtual tours.

Neither Foxtons nor its suppliers will release cost details for such refurbishments. But given that the company's chief executive, John Hunt, features annually in *The Sunday Times* list of the richest 500 individuals in Britain, you can judge for yourself whether the innovation generates enough business to cover the cost.

New media

Interactive television services

These are in their infancy but already there is a UK television channel dedicated to selling homes, advertising them in one-minute slots that can be seen by subscribers to Sky Digital.

TheMoveChannel.TV is a spin off from *www.themovechannel.com*, a website which primarily advertises homes for sale but offers high volumes of information on buying homes in the UK and overseas, and news on the property market.

The TV service features a rolling 15-hour sequence of video advertisements for homes on sale, with each hour dedicated to a different region. Property details are shown as on-screen text looking a little like the BBC's Ceefax service, displayed with the selling agent's or developer's logo and contact details. Viewers can use the interactive facilities and an accompanying special website to request more information from the agent or developer via the internet.

The website includes pages of estate agent data containing details of their services, e-mail addresses and links to the agent's own internet service; the site also carries property listings, featuring basic information on each home shown on the TV channel with a thumbnail image and a short description. The site also carries links to other information on the internet, including sponsored micro sites dedicated to services such as mortgages, solicitors, surveyors and removal firms.

"This will improve the buying process for consumers, allowing them to filter out unsuitable properties and spend less time on wasted viewings, with the result that the leads agents receive will be of an exceptionally high quality," was how the network's founder, Dan Johnson, described it at the time of its launch.

TheMoveChannel's managing director, Kevin Cochran — who has worked with several digital TV projects unrelated to residential property — says: "We've learned from the mistakes made by some digital TV companies, which have extensive studio time, high overheads and, as a result, are too expensive".

Not everyone in the industry is as enthusiastic and the channel's launch in 2003 was delayed because estate agents had been reluctant to embrace the concept of TV advertisements. Whether those opposed to it were simply reluctant to embrace a new idea or genuinely believing it may not enhance sales is up for debate.

"It isn't immediate enough for today's busy home hunters. Who has the time or the inclination to sit for an hour and look through a number of properties that may or may not fit their criteria?," asks Mark Howe, group marketing director for Bradford and Bingley estate agents, which decided not to use the new service. "If you could see a number of properties on a screen, then click to the one you like and watch a short film, that would be one thing — but not watch video after video.

"Advertising properties through TV isn't viable. Profit margins are very tight within estate agency and TV advertising is expensive. They aren't a good match. We've never been involved in TV advertising on the estate agency side and cannot seeing it being a future option," he added.

Neil Mackwood, former head of property website *www. 08004homes.com*, which closed in 2001, says the idea is "a dead duck". *www.08004homes.com* used to run some of its properties for sale on two London cable TV stations. One of them also featured extensive advertising from London agents such as John D Wood, but Mackwood says feedback was poor. "Those were the early days for interactivity and digital TV, so things will be better now. But the punter is probably better-served by using a large website and looking for a property with the immediacy and detail that the Internet offers," he suggests.

The new channel can draw some comfort from what happens overseas.

US estate agents commonly advertise on television, although they usually promote their general services rather than individual homes for sale. But some now buy TV space to show off exceptional properties, featuring video clips of each room and the garden in 45-second advertisements.

UK house builders who specialise in investment properties in prime central London sites often advertise their new apartments in Hong Kong to encourage investors from Asia to buy.

TheMoveChannel's Kevin Cochrane remains optimistic but believes the entrenched views of what he calls "conservative minds" in the UK estate agency industry have to be won over.

"The property world has to wake up to the reality and strategic possibilities of digital TV. There's a need for developers in particular to differentiate their product and strike an advantage over rivals," he says. "This will create a much greater impact for a property and a brand than having a large presence on the Internet. TV is the future."

Telephones, especially mobile text messaging

The UK has over 40 million mobile telephones and approximately 30 billion text messages are sent from and to them each month. Where once texting was the preserve of the teenager avoiding voice calls to save money, the phenomenon is now a communications medium in its own right and is espoused by people of all ages.

Software packages exist that allow text to be written on to a pro forma on a personal computer, just as an e-mail may be written. But instead of being sent to another single user, it can be "transmitted" to hundreds or even thousands of people's mobile telephones at the same time — or, if the client is special enough and your resources appropriate for such one-to-one treatment, it could be sent to just a single recipient. The point is that it is easy and does not even require the dexterity of texting from a telephone keypad.

Most estate agency software being developed in this medium allows agents to improve communications with clients and their own staff, and allows an office to:

- send suitable applicants details of a new instruction within seconds of it being agreed
- send vendors confirmations regarding appointments, viewings, offers, or completion and exchange information
- send appointment reminders to negotiators and valuers out on the road
- send ad hoc messages as appropriate.

A good example of this still-developing facility for agents comes from Estates IT, a software provider well known within the property industry. Using its management software package called PCHomes, its promotional material illustrates how relatively inexpensive the facility is and how it can generate savings of time and money.

First, it claims that a branch office with a telephone bill of £250 per month may spend about 75% of that on mobile calls (bearing in mind the cost of calls to mobiles are much higher than those to landlines). In other words, about 300 mobile calls each at an average cost of 60p plus VAT. It claims that it can offer mobile telephone text messages at a cost of 7.5p plus VAT each, saving 52.5p a call and therefore a saving to the branch of approximately £160 per month. Even if text messages replaced only 50% of mobile calls, there would obviously still be a saving of £80 per month, or almost £1000 per year.

Second, it sets out a possible saving if a new property coming on to an agent's books fits the bill for a large number of clients. Estates IT's scenario claims that if, say, 40 applicants were interested in a particular type of property it would cost £24 plus VAT and two hours of time from a member of staff to contact them on their mobiles — or longer if they have to give extensive details of the home that had just been put on sale. A text message sent simultaneously to all 40 would-

There are over 45 million mobile telephones in the UK. Text messaging is becoming more common — estate agents alert buyers to homes just coming to the market, and let sellers know of new offers or feedback from viewings. The new generation of mobiles send and receive pictures — Vebra, among other firms, is experimenting with this technology.
Source: *Vebra*

be buyers would cost only £3 plus VAT and could be done in as little as one minute.

There is one other attraction that Estates IT and other software companies offering these services curiously omit from their publicity — that is, that members of the public (especially younger ones in the first-time buyer category) prefer text messages to mobile or landline calls. Even the promise of offering such a facility may attract business to an agency and give an aura of modern professionalism.

Smart Phones

The BlackBerry is the best-known example of the so-called Smart Phones, which combine mobile telephony with PDAs, internet access, a digital camera, e-mail and even MP3 music players to boot. They have a traditional QWERTY keyboard so can quickly become familiar to anyone accustomed to using a personal computer.

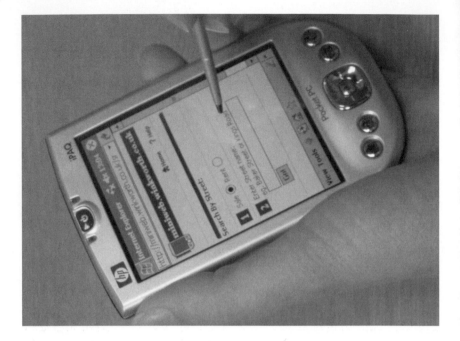

Winkworth, a London estate agency chain, is sending a "simple" version of its website to hand-held Personal Digital Assistants and users of so-called SmartPhones which can surf the internet. Other agents are likely to do the same within the next two years.
Source: *Winkworth*

One UK property data service, Property Intelligence, is marketing a modified form of the BlackBerry, which can read a continuously updated catalogue of information. The company holds details on more than 400,000 commercial buildings, including addresses, owners, occupiers, deals done, space availability, and high-resolution photographs and sells it to commercial property professionals in a form that can be read on these hand-held devices.

The Property Intelligence version is tri-band, so users travelling to the US can use it to check e-mail as well as read the documents.

So far, there is no comparable software for residential agents to use for would-be home buyers with BlackBerry telephones — but who would bet against one being created within a matter of months?

Case Studies of Adventurous UK Property Pioneers

The UK general public are by no means backwards when it comes to using new technology, as the prolific use of mobile telephones and the increased use of remote, wireless laptop internet services by many business people illustrate.

But it is clear in many industries that if firms are doing well, operating at what many believe to be an adequate level, there is little incentive to modernise. This seems to be the approach of the residential estate agency industry — it has seen the widespread introduction of internet sites over the past 10 years, so why does it needs more technology?

There are, of course, notable exceptions to that philosophy and some of them are featured in this chapter.

But there are also some inclusions here of non-estate agents who are using this technology to undermine the traditional estate agent as well as some using it to bolster the cause. Both sides have been included because, unless the industry as a whole re-equips itself, I believe estate agency is likely to fall victim to more of these dilettante operators who think technology can allow homes to be sold on the cheap, ignoring the expertise that the residential industry has built up over the past half century.

Case study 1: Virtual estate agencies — Charles Lister, Daniel James and My First Home

I'm trying to fight against the paranoia of the existing estate agency industry in the UK, which has individual agents preoccupied with trying

Just another estate agency website? Well yes ... and no. Charles Lister is one of the UK's first "virtual agencies". It has no physical office at all, and sells properties soley through the internet. The site is more sophisticated than most — note the clever map functions.
Source: *Charles Lister*

not to lose any instruction to any competitor. It's mad and, ironically, it's keeping down their fees.

Are those the words of an outside consultant looking in at the industry? Or a consumers' champion taking a simplistic critical position against estate agents?

No. These words are from Henry Pryor, an estate agent formerly employed by Savills, Strutt and Parker and The London Office. His outburst came as he opened a "virtual" estate agency called Charles Lister that has no physical office and is using technology to cut costs, improve service to customers — and start a campaign to increase those fees.

Charles Lister works like this.

- First, it recruits estate agents to subscribe to the company at £1000 per office per year, or at a rate of £100 per office per month.
- Second, it then uploads any or all of the subscribing agents' property portfolios from their own databases and websites on to

the Charles Lister site (*www.charleslister.co.uk*). The total contents of the site are accessible only to subscribing estate agents but members of the public can interrogate the system as they can any more conventional estate agent's website.

* Third, any one agent — even if he has properties from only a small geographical area — can therefore respond to a request if a customer walks into an office in, say, central London and asks if there are properties available in, say, Yorkshire. The agent simply scours the full database for a suitable home.

Using the internet, Charles Lister links over 150 individual agency offices across the UK. It charges a seller a "joint agency" fee of 2.5%, payable only if the property is sold as a result of being seen on the site. The fee is shared between Pryor and the relevant "physical" estate agency office that is marketing the property.

According to Pryor, the seller gains because, instead of paying one estate agent office as little as possible (usually 1.5%) to sell their house, an extra commission of 1% or so allows the property to be linked to hundreds of others. "As a result, our clients get the best professional advice on the marketing of their property with the knowledge that we are striving to get the best deal regardless of where the buyer is found," he says.

Pryor spent some £60,000 on setting up the firm — a hefty contrast to the typical £120,000 that would have to be spent on the first year's activities of an office with two staff in an out-of-London town.

This is a very small form of the Multi-Listing Service that is operated widely in the US and in some parts of mainland Europe too. The difference is, in those countries, agents charge sellers a hefty commission fee of between 4.5% and 6% of the agreed sale price.

Pryor is keen that the website should give the impression that the firm is a "real" estate agency. Prospective home buyers can feed in the usual criteria of price and house size, and an onscreen mapping service allows a very precise location to be chosen if the would-be purchaser wants to be more specific than the usual insertion of a town name or larger postal area.

Charles Lister also boasts membership of the Ombudsman for Estate Agents' scheme and puts its property portfolio — which is an aggregation of many agents' portfolios — on the Primelocation property portal, just as a conventional agent would.

"I have no shop window premises. What is the point of having an expensive office in somewhere like central London if I am trying to match buyers and sellers in two completely different parts of the country?" Pryor asks.

"And why should I employ expensive negotiators to sit around offices while their expensively branded Mini Coopers sit idle collecting parking tickets outside, especially if the market is slow at any time?"

Of course, there are many political obstacles to Pryor's objective of kick-starting multi-listing operations in the UK. For example, high street estate agent chains that already have offices across much of the country would probably see little advantage in adding their extensive portfolios to this scheme, and if they were to do so they may inadvertently give business to a rival agent who is spotted by a user.

Likewise, the large scale-existing property portals, such as Rightmove, Assertahome and Fish4Homes, may claim that their contents of tens or even hundreds of thousands of properties already provide what is effectively a multi-listing service without the need for individual agents to subscribe to a special club such as Pryor's.

It is not the job of this book to argue for or against such a scheme, which in many ways is merely a variation of the "Solicitors Property Shop" concept that exists in most Scottish cities and allows the public to view a range of portfolios from different selling companies. But if agents throughout the rest of the UK wanted it, the technology exists to support it.

By using the standard property card layout available on most of the estate agency software, the same property details that go to make up the database in an agent's office can be accessed by other agents if that is required, and can provide automatic content for an additional hybrid website such as Charles Lister.

"We started off in September 2004 with a critical mass of 3500 properties and hope to keep getting agents on board. Currently, they probably couldn't justify charging more than the market norms for commission," admits Pryor.

"But in Cirencester there are 24 estate agents' offices. In Battersea, there are 22. Today, if a buyer wants to get a home in one of these locations he has to visit every single agent and register separately with each one. It's a huge chore. If a seller can get access to all buyers signed up to every agent by simply registering with one agent signed up to a multi-list system, then commission could rise justifiably," he argues.

"As with all things technical and innovative, however, the property establishment's lethargy may yet kill it off!"

Within a few days of Pryor saying those words, the backlash from some parts of the property establishment began.

Melfyn Williams, a former president of the NAEA and a director of Anglesey-based agency Williams & Goodwin, said: "People keep trying to bring in multi-listing because it works in America, but the big difference is that they charge 6% commission and we only get 1.5%."

A more modest example of the virtual agency concept can be seen on *www.daniel-james.co.uk*, the website of a new Suffolk estate agent.

It is much smarter than the usual small, local agency website and has the usual search facilities for properties defined by price, size and location. Yet is there something else about it that is unusual?

Only when you hit the "contact us" link on the home page does the secret creep out — the postal address for Daniel James Estate Agent is a post office box instead of a high street office in its catchment area of Haverhill.

James Rogers set the business up in spring 2004 deliberately wanting to avoid having a shop front office for three reasons.

First, by relying on the Internet, leafleting and extensive local press advertising he felt he would enjoy a quicker and cheaper business start-up than having to find a good location and refurbish an office with a traditional shop front. Second, his much-reduced overheads could be passed on to customers directly through a lower commission and, he hoped, consequently larger volumes of business. Third, the risk to himself and other investors in the new firm would be reduced.

He spent about £10,000 on software and hardware, which are used to almost simultaneously do three things.

They provide internet advertising of his vendors' properties (sent directly to his own site and then the Rightmove and Vebra portals to which he subscribes); the software then produces the conventional hard-copy details to be mailed to would-be buyers who have registered through the website; and finally the software submits the details to form ads in the property supplements of the local press. "The pictures and virtual tours we do ourselves, using Vebra software to add them to the site and details," he says.

His website contains downloadable floor plans and general property details too, and interested parties can register online to view a property. All the information is organised by Rogers and his colleagues at a private office hired cheaply in a non-prime area.

"It's been a success. Of course, this isn't the right approach for everyone — we've had a few members of the public who haven't instructed us because they're worried about not being able to go into an office on a street. But we've had other instructions directly because of the lower fees and because people recognise most home buyers now start their hunt on the internet," says Rogers, who worked for other corporate agents in traditional offices in the area before setting up Daniel James.

Rogers insists that the focus of the business remains local, even though being internet-based means he could — in theory — stretch the catchment area for clients beyond the geographical boundaries that a traditional office might be forced to accept.

As with the Charles Lister service, Rogers' agency goes to some lengths to stress it is a "real" house-selling business that happens to use technology, instead of being an internet company or a virtual office that happens to sell homes. Its trained negotiators make pitches, offer valuations and accompany viewings as is the case with most traditional agents.

Being an electronic service has not prevented it from joining the NAEA or subscribing to the Ombudsman for Estate Agents scheme. The website also gives details of how people can fax, telephone, e-mail or write to the firm — the only old-fashioned action they cannot do is walk in to a branch office.

But this statement on the website gives a flavour of the firm's technology-rich approach:

> Complemented with traditional estate agent methods, we use the very latest computerised technology, to ensure that each property is marketed to its fullness, to ensure the best possible price is achieved. Each property will enjoy maximum exposure, being displayed on the internet with downloadable colour details and 360-degree virtual tours, as well as colour advertising in the local and regional press.

> A personal service
> * Competitive fees.
> * Fully trained and professional staff.
> * Accompanied viewings, when required.
> * Marketing to suit the vendor's needs.
> * Full sales progression and vendor reports.

> The latest computer technology
> * Full-colour digital property details.
> * Floor plans.
> * 360-degree virtual tours.
> * Full colour advertising in local publications.
> * E-mail alert mailing list.
> * Multi-website advertising, including *www.rightmove.co.uk*, *www.vebra.co.uk* and *www.daniel-james.co.uk*.

The unique selling point that Rogers feels he can offer the public as a result of all this technology is a service that is at least as good as any from a traditional office-based similarly sized agent but with much lower charges — usually £1000 plus VAT on a no sale, no fee basis.

The Daniel James agency was launched in a period when competition was strong between local agents in the area, and when the number of transactions was reducing slightly because of increased interest rates taking the edge off the market.

"My competitors have vastly greater overheads than I do. I certainly do not want any one of them to fail but if we're in a more difficult market situation I really do believe my business has the advantage of being able to offer the same service at substantially lower costs, both to my clients and to me as a manager," says Rogers.

He says the concept has been an immediate success well beyond the expectations of his business plan. He even has some clients who are older and not computer-literate, but have accepted the cost and penetration advantages of his sales approach.

"It's not that hard to see it makes sense," he says.

A similar venture is under way with *www.myfirsthome.co.uk*, which is another website-only service. The firm is based in Worcester but has no office that is open to the public although, again, it operates similar to a conventional agency in most other ways.

It also takes a braver, entrepreneurial approach to fees.

It includes the conveyancing charges for the sale of freehold properties within the standard fee of £1250 plus VAT for properties worth £175,000 or less. This is possible by the firm having "pre-booked" conveyancing services with a local solicitor.

Alternatively, clients can pay £850 plus VAT if they wish to use their own solicitors. In both options, the increased complexity of leasehold properties means slightly higher fees.

As with the Daniel James example, the My First Home service is winning plaudits from the public.

"I think that eventually high street estate agents simply won't exist in the way they do today — there won't be the need. People will be able to download all the information they need off the internet and then arrange a viewing. It's much simpler than trudging around different offices," according to My First Home manager Sam Nall.

Case study 2: Hand-held technology — Winkworth's Miniweb

This estate agency has 60 offices across London, the south east of England and Yorkshire, and has launched what it calls Miniweb. This is a so-called "lite" version of the company's conventional website, which has 200,000 hits each month.

It is specifically for viewing on hand-held devices, such as PDAs, so-called smart phones and BlackBerry devices — in effect, anything portable as long as it has an HTML browser that can download the internet.

This means that the property search features from Winkworth's main website (which in an unchanged format would be too data-rich and dense to download to the limited memory and screen of a hand-held device) can now be stripped of unnecessary data. Only "bare bones" internet pages are downloaded to the portable devices.

Hand-held devices suffer slow connection speeds, have small screens and can use internet browsers with only limited capabilities; therefore, they cannot support the download of standard websites. The Winkworth Miniweb is predominantly text-based and contains few images, so those problems are minimised.

Unlike the traditional PC, hand-held devices do not have the luxury of a mouse or keyboard. Instead, PDAs use a stylus, a small pen-like instrument, which touches different parts of the screen to trigger new pages or to offer more details.

The Miniweb's "Drill-down Wizard" was designed to make property searches compatible with these interfaces. It uses single clicks or taps to fill the required fields prior to commencing a property search.

Winkworth was one of the first estate agencies to launch its own fully interactive property-related website with the function to carry out a comprehensive property search. www.*winkworth.co.uk* was also the first property website to introduce virtual tours.

Case study 3: The extranet — a "secret" service that can be offered by agents

In summer 2001, Savills, or FPDSavills as it was then, publicly stated that its newly relaunched website should change the way that their customers related to the firm.

One of the catalysts for this was the creation of a password-protected extranet that allowed the confidential exchange of information between agent and client.

An extranet is what the computer world calls a private network that uses internet technology to share confidential business information between external suppliers, vendors or buyers. By the nature of the data they carry, extranets are usually encrypted so that only the user or users with the right decoding software can access the data, which may also be password protected for extra security.

In the Savills example, a vendor could sit in their other home in Cape Town, or even an internet cafe in New York, and use a unique password to gain confidential access to "their" part of the extranet to check floor plans or written particulars before they went into a sales brochure or on to the public area of the website.

Savills is a particularly wealthy agency and, because of its global reach, has no fewer than 31 staff in its IT department. At the other end of the scale is Waterfall, Durrant and Barclay, a three-office agency in the Woking area that uses a variation on the intranet to allow vendors to track the progress on selling properties online.

Intranet contents include buyers' feedback after viewings, and the number of enquiries about the properties received in the firms' offices. The company even has a customer liaison officer dedicated solely to handling online activities, freeing traditional negotiators to do their core work.

The company claims the approach has won it more business and allows easier exchanges of information — and the initiative was given the title of Best Use of Technology at the 2004 Estate Agency Of The Year Awards.

Extranets are simple and inexpensive add-ons to internet sites and can be created by even a modest local agent. Yet relatively few estate agencies have followed Savills' and Waterfall, Durrant and Barclay's example, although those that have include upmarket central London firm Kay & Co, the Knightsbridge property consultancy Holmans, and the Warren Rowley Hardman agency in Cheshire.

But an extranet can offer a secure route to exchange information when the client and the agent's office are in different places and, especially, in different countries operating in different time zones. Why make a midnight telephone call to catch a client in Australia when a message left on a secure extranet, complete with plans or maps or photographs, can be viewed by the client in his or her own time zone, with a prompt reply ready for you when work starts at the office next morning?

Case study 4: Small business but big technology — Northumbria & Cumbria Estates, Hexham

It is little wonder that Northumbria & Cumbria Estates (NCE) was awarded the title of the UK's Best Independent Estate Agent of 2004 — and it is not coincidental that it puts its success down to new technology.

The firm started in 1989 with three staff and ungainly premises on the first floor of a building in the town centre. "Everything was manual — we relied heavily on telemarketing to generate business; property and customer details were handwritten on to cards; and mail-outs were done by a local printer then put into envelopes by our staff," proprietor Andrew Coulson explains.

Today, the firm does not use paper. It puts customer and property details into a PC database. It alerts house hunters by mobile telephone text and e-mail as soon as an appropriate property becomes available, and it e-mails digital photos to keen clients. It only sends out hard-copy details if vendors or would-be buyers request them.

The company has even made a DVD about the region, which it proposes to make available on the internet to lure buyers from elsewhere in the UK and overseas, and it has sponsored a webcam project that puts non-stop live pictures of Hexham on the web to promote the area.

The firm says this investment not been an idle exercise in playing with gadgets. Revenues have grown by 300% and the firm claims to have sold a home in Wales to a buyer who did not visit the property but relied on digital photographs viewed over the internet.

NCE invested £11,000 in a software package in 2002 and pays an annual maintenance fee of £1500 per year. BT charged the firm up to £1600 per quarter for telephone calls and faxes before the introduction of broadband — it now pays about £1500 per quarter but has much quicker access to surfing the internet and updating its own website on *www.northumbria-cumbria.co.uk*.

Winning the award and using the technology has given it good publicity too. "As a result, sales have rocketed and we have exceeded all our business targets," the firm told *The Guardian* at the end of 2004.

Case study 5: Internet property auctions — slowly but surely they are working

The recent history of property auctions is reminiscent of the heroic failures that are the stuff of British culture — we prefer a plucky athlete coming second after a struggle than a dominating champion who effortlessly finishes first, or a gallant actor nominated for an Oscar than one who easily wins it year after year.

In the cases of three property auction sites launched between 2000 and 2003 we certainly have failures but also a taste of what the future will hold — and a much more recent example which had more success.

Internet property auctions are little known because the few sites that exist offer only informal, non-binding arrangements that look casual to many property professionals. In addition, the sites have not tried to go beyond low-cost, residential properties and do not attract upmarket houses, nor investment and commercial properties.

The site that used to exist as *www.propwatch.com* (do not worry about looking for it today as it has been taken over by a North American advertising group) failed to sell a single property in a year of its existence as a UK auction site. Although it had an anonymous, experienced auctioneer among its founding members, it failed to get other property professionals involved, which may have been its undoing.

But perhaps the most obvious cause of its failure was that it never attracted a critical mass of properties to sell. After six months of existence, it had only 40 properties from across the UK registered on its site — by that time, some of those had been there for many months and even the site's owners believed the majority may have already been sold through traditional estate agents.

According to Naresh Sonpar, founder and chief executive of Propwatch during its short existence, the fault was neither in the technology nor in the principle, but solely in the small marketing budget created to launch the concept. "We were a tiny company and we just couldn't keep up with the likes of Savills or others, which could set up a site and sell it to the public," he says.

Another online auction house, *www.thepropertylot.com* (where again the site no longer exists on the internet), did not follow Propwatch by trying to get people to use the internet instead of using an estate agent, but actively promoted a twin-track approach.

"Auctions are no substitute for the role of the local agent," according to Martin Constable, an IT expert and founding director of Thepropertylot. His business started off mainly with investment and buy-to-let properties, some of which were advertised also by housing associations. He planned to go to local estate agents in the areas where the properties were located and ask them to handle viewings and on-the-spot enquiries.

The idea did not take off — again, there were too few properties advertised to make it a strong or attractive proposition to buyers (let alone sellers, who could see the paucity of other properties on sale), although it was a lack of co-operation by local agents that was officially blamed for the site's demise.

However, its principles are ones that could be taken up by estate agents themselves if they were sufficiently entrepreneurial. Only a fool would believe something similar would not re-appear, so an analysis of how *www.thepropertylot.com* operated may provide food for thought:

- Properties were put to auction, with details on the site, with no entry charges for sellers.
- An online catalogue, to be available about three weeks before each auction date, would include many more sales particulars, including a range of photographs, detailed floor plans and any special conditions of sale.
- Local agents would be approached to handle physical viewings of properties, with "physical" viewing times specified on the website; this would be in addition to a 24-hours-a-day online viewing period for about two weeks before the bidding commenced.
- The site offered a downloadable legal pack, available at a charge, to help with conveyancing if anyone wanted to do that themselves.
- Each property had its own timetable for sale, normally in the form of a single 24-hour online period.
- Bidders' identities were confidential, but the size of their bids were to be made visible — of course, this was to encourage other, higher bids

In practice, a major deterrent to its success was the fact that to become a bidder, a visitor to the website had to register an interest in a specific property — and the site's operators insisted that this meant signing a cheque for 10% of the guide price of the property the individual was interested in buying. The cheque was not cashed unless the bid was successful, but this was (perhaps rightly) a huge obstacle. Members of the public in 2001 and 2002, when this site existed, were even more sceptical than today of websites seeking cheques to be sent to strangers.

Constable says this approach was necessary to prove good faith on the part of the would-be bidder, and to give time for Thepropertylot to test the financial bona fides of registered applicants. Nonetheless, it helped guarantee the site's ultimate failure.

The third attempt at a property auction website was undertaken by developer Sunley Homes, a firm known for its innovative residential developments and also its bid to encourage online auctions for new homes.

As far back as 2000, it experimented with putting two new apartments in one of its developments in Eastbourne on to a specially created website called *www.homes4living.co.uk* (which, you may have already guessed, no longer exists).

Interested parties had to register, without any pre-payment, to receive details of how to visit the properties prior to bidding. But, from the outset, the scheme had mixed success and only 300 people did actually register out of an impressive 150,000 people who visited the website over a two-week period.

But only a disappointingly small number of people visited the development in person and the online bids that came in — Sunley will not reveal the number nor the price the apartments were eventually sold at — were probably much the same as might have been expected had the properties been sold through a traditional route.

"Technically, it worked perfectly but people were just reluctant to commit online," says Sunley's John Keele.

The company talked to other developers at the time but they all believed the internet needed more growth — but in the few years that have elapsed, the proliferation of users, exponential growth of web traffic, and general familiarity with online shopping means that the time may now be ripe to revisit online property auctions.

Evidence that this is the case came in late 2004, when Applecross, a small developer in Scotland, decided to sell the first phase of its 24-unit development at Eyre Place, central Edinburgh via an online auction.

Bids were to start at the reserve price of £420,000 for the three-bedroom apartments, although if buyers put in a significantly larger bid at the outset they could ensure the property of their choice was removed from the bidding process.

Potential buyers were required to register through their solicitors in advance to participate in the bidding process and the close of the auction was staggered at 15-minute intervals to enable buyers to bid on alternative apartments if they lost out on their "favourite" property.

In the end, only two of the eight flats were bought this way, although they included one taken by a Far East businessman who had never even visited the city. However, although the reserve price on each flat was £420,000, Applecross says the two sales netted the firm over £1 million, indicating that successful online auctions can encourage bidding at well over the asking price.

The world's most successful auction website, eBay, has not been immune to the residential property world. Users register for a minimal price, slot in one or more digital pictures of what they want to sell, state a reserve price below which they will not sell, and suggest a duration for online auction bidders — usually one or two weeks. The highest bidder at that time buys, providing the sum exceeds the reserve, with financial dealings usually through a secure electronic third-party operator that checks the financial bona fides of both seller and buyer.

The residential property section of the UK eBay site lists around 60 properties at any one time, a figure no doubt set to rise in future.

Louise and Robert Layton put their Cornish farmhouse and its equestrian centre and an accompanying 14-acre plot of land on eBay in November 2004, with a starting price of just under £1 million and a 10-day auction duration.

"A house at this price will probably be bought by someone from outside of Cornwall because local wages here aren't very high. So we want people around the country to see it and eBay is a cheap and effective way," explains Louise, who ran a livery stable at the property.

Husband Robert, the owner of a charter diving business, regularly bought and sold on eBay and was encouraged to market the house this way when he purchased a car through the site without even seeing it in person.

The farmhouse cost just under £40 to list on eBay (the fee being much higher than most associated with eBay, because of the high value given to the property). They described it as "a wonderful little piece of Cornwall" and gave elaborate details of the accommodation along with several pictures.

Now internet "sales consultants" are springing up to try to improve the presentation of properties listed on eBay and other auction sites.

Christian Braun is managing director of Auctioning4U, a firm that takes professional photographs of properties that sellers want to advertise on eBay. It also completes the extensive online forms that have to be submitted by sellers. He charges a £250 listing fee for a property, plus a further £500 if it sells.

"People must remember that the first nine days of a 10-day eBay auction are just for looking or making enquiries with sellers, so bids at that stage always appear low. The last hour and final few minutes see the very competitive bidding that can create high prices if something is promoted properly," he claims.

The Laytons, who did not use any sales consultancy, received over 1,000 enquiries for their eBay auction but no bids above the reserve price.

Yes, it was ultimately a failure — but who would bet on this not being a successful sales route in years to come?

Case study 6: *www.reservethathouse.com* — another failure but another warning to agents too?

www.reservethathouse.com was an online property idea that tried to paraphrase John Lennon in putting across the message: "Imagine there's no estate agency — around us only homes". It also tried to do for house purchasing what friends reunited (*www.friendsreunited.co.uk*) has done for school and workplace reunions.

The site worked like this:

- You saw the ideal property or street in which you want to live. You registered your own, current home on the reservethathouse website, and then listed the name, number or postcode of the house of your dreams that you spotted earlier. Over time, you could see if anyone had shown an interest in wanting to live in your property — although you saw only the expression of interest, not the name or contact details of the enquirer. So far so good, and it was all for free.
- The second phase cost just £5. This allowed you to see the contact details of the people who had registered and expressed an interest in your home. It also allowed you to send one pro forma letter (written and mailed by the operators of the website) to the owner of the home you had your eye on. If you wanted to contact the owner of more than one property, you paid £2.50 for each subsequent letter.
- Finally, to save you checking back to the site, any response to your letters and any further interest in your own home was sent to you by e-mail.

And that was it — a simple idea which, at first sight, seems naive, whimsical and entirely without hope of success.

But think again. In the first 18 weeks of its existence in 2003, and despite a near total lack of publicity, it received over 1000 entries per week. After six months the conversion rate (that is, the proportion of registered users who sent one or more letters to existing homeowners) was around 20%. Then, after a lengthy feature on the site in the popular *Steve Wright In The Afternoon* show on BBC Radio 2, registrations rocketed and the conversion rate for the next six months hit 30%.

A house hunter in Nottingham mailed the owners of all but one of the houses in a whole street in the city; another would-be buyer contacted the owners of over 30 homes on the Isle of Arran.

The site no longer exists suggesting that interest waned, the novelty wore off and buyers saw the naivety of their ways. But there is a good chance that something similar will be set up in future — and could estate agents learn anything from this aggressive venture?

www.reservethathouse.com wanted to use technology to change the way Britons viewed house buying and employed populist measures to play on the public's desires to move to certain streets and localities because their reputation was high or because the nearby schools performed well in league tables.

"We were very keen that people stuck with professional advice all along the route except for estate agents," according to Dave Warriner, one of six entrepreneurs behind the site. He described himself "not anti-estate agent, just pro people".

"We advised people to get their own property valued by a chartered surveyor and not to upset agents by getting them to do it and then not instructing them. Likewise, if people wanted to sell, they shouldn't do their own conveyancing but instead use solicitors," he says.

So what was so wrong with estate agents?

"Well ..." he hesitates, seeking the right euphemism. "We were not trying to do what they do. They sell houses that people want to sell. We were trying to help people buy the houses they wanted, even if the owners didn't want to sell at that moment."

The letters that the site sent to owners explained that someone liked their home and that even if there was no intention of selling it now, it might be worth keeping the contact details of the interested person and getting in touch at a later date.

"If that eventually led to a sale, then great. Frankly, it would have saved them paying 1.75% of the sale cost to an estate agent, so they would have benefited," claims Warriner.

Reservethathouse only ever sent about 7000 letters — although Warriner adamantly reminds people that *www.friendsreunited.co.uk* existed for much of its first year with only 3000 people registered until the right publicity triggered a deluge of applicants, which helped it become a global internet phenomenon, spawning TV game shows and fly-on-the-wall reunion documentaries and newspaper articles.

Most estate agents have had some experience of "cold contact" selling, so are there lessons from reservethathouse for the future?

The Pimlico office of London estate agents Douglas & Gordon has contacted all of the owners in one street after a would-be buyer said he wanted an unmodernised house there — and it worked.

Upmarket London agency Farrar has helped a buyer who narrowly missed out on buying two flats through the traditional route. "Knowing that one of the flats had been sold as an investment, I put my business card through the door asking if the new owner might consider selling," recalls Farrar agent James Pace. "The owner agreed, making himself a tidy profit of £40,000 in the process and avoided paying agent's fees as I was acting for the buyer."

How straightforward it would be to do more of this using new technology to target the right sort of home owner in the right sort of property in the right sort of street — all you need is the right sort of database and a standard letter.

www.reservethathouse.com may have been an anti-estate agent website that has now bitten the virtual dust, but its lasting legacy could be that it alerted the people it liked least to the potential of pro-active marketing thanks to new technology.

Case study 7: Mobility-friendly homes — using technology to create a niche market

Wheelchair-friendly doorways, sturdy handrails and nearby car parking: these are what the UK's 8.3 million registered disabled people and those with mobility problems look for in a home, but rarely do they appear in an estate agent's particulars for any property.

But now shrewd use of the internet has created a useful niche market for an agent and may be prompting the industry to rethink this social issue.

Mike Reid, a founder of Sussex chartered surveyor practice Reid+Dean, has set up *www.mobilityfriendlyhomes.co.uk* to encourage agents across the UK to publicise residential properties that are for sale and have been modified with handrails, ramps and wider corridors and doorways.

"I worked with local authorities assessing public buildings for access to conform with the Disability Discrimination Act but realised that no one was doing the same for residential. The idea grew from there," says Reid, who works on commercial property valuations, rent reviews and lease renewals on top of his own residential agency business.

He says there are hundreds of thousands of modified properties for sale but buyers cannot know because these features are usually excluded from conventional agents' descriptions.

The mobility website encourages agents to accept commissions from properties that are particularly accessible and urges them to list their characteristics, such as nearby unrestricted same-level car parking, entrance areas that could have a ramp, and a toilet at entrance level.

The Part M building regulations, introduced in 2000, mean that new-build homes constructed since that time must have at least basic access for those with disabilities, but that leaves more than 97% of the UK housing stock with potential problems for such users.

The Mobility Friendly Homes website is modest in technical terms. It is modelled on Reid's own mainstream residential agency site and offers general descriptions of houses and flats on sale across 13 regions of the UK. There are also regular search, contact and "about us" components. But a glimpse at any specific property details and you will see why the site has been established.

The details of a home may look familiar until you read that it has "a fitted stair lift" and that "a ramped access could be created to the front and rear" while "a few outside steps with a handrail lead to the front door". In other words, the adaptations are overtly included in the details instead of being omitted or worded in a way that would not deter able-bodied clients.

The advantage of modern software is that the same property could, in theory, easily appear on two websites — one with the euphemistic wording to appeal to the widest audience, while another more explicitly worded version could appeal to specialist would-be buyers.

"We've no real budget so we rely on individual approaches to agents and on people who are selling an adapted home. We'll then contact their selling agent to get involved in the site," explains Reid, who, in late 2004, started promoting modified rental properties on the site too.

Mainstream estate agents have to apply for a licence to allow them to use the site to publicise modified properties or to use the *www.mobilityfriendlyhomes.co.uk* logo and link on their own websites. They can also use its branding in their offices or on their letterheads, too, if they are licensed.

The site also allows individuals to market adapted homes privately for a token fee of around £25 for three months' promotion.

It is tempting to see this as a worthy but ultimately uncommercial exercise but Reid thinks otherwise. He points to the 8.3 million registered disabled people in the UK and to the fact that, by 2020, more than 40% of the population will be aged 50 or more and, therefore, increasingly vulnerable to mobility difficulties.

Reid also says the bottom line for agents and their commission might benefit from a more positive approach to this previously taboo subject.

"The technology allows us to do this where if we had tried to set up a separate office dealing wholly with this specialist market, it wouldn't have made sense," says Reid. "It's a real winner."

Case study 8: House Network — the shape of things to come (without estate agents)?

How much does it cost to move house?

On average in 2004, for someone moving from a home worth £150,000 to another property valued at £200,000, it cost over £5200, according to house moving research data updated annually by the Woolwich building society and the University of Greenwich in south London. That is a UK average — if the move is from, to, or within Greater London, it costs up to 25% more.

Much of that cost across the country is down to stamp duty and almost unavoidable search and legal fees, but, on average, £2553 of those fees go to estate agents, according to the research. We in the industry know this is little compared with buying similarly priced properties overseas but the perception to the UK buyer is that estate agency fees are high and sometimes undeserved.

As a result, there are growing numbers of house-selling websites that seek to bypass estate agents. At the moment, these do not represent a significant threat to the industry but, one day, they might, so it is worth observing their techniques.

Among the newest and best publicised is House Network (*www.housenetwork.co.uk*), which claims to provide "advanced, automated marketing technology and friendly support staff," allowing members of the public to sell their homes privately from just £49.00, with a more sophisticated package costing only £115.

House Network's publicity makes it clear that it is playing on anti-estate agent sentiment.

> A typical agent will charge you 2%. We have no hidden charges whatsoever. Ask yourself the question "What does my estate agent do to justify the fee they charge for selling my home?" You'll soon run out of answers and wonder why you have used them to sell previous homes for you! The concept is quite simple, the only additional "work" that you have to do is as follows — and we're sure you'll agree that it's well worth the thousands of pounds that you'll save:
>
> 1.   Measure the rooms in your home.
> 2.   Take a photo(s) of your home and upload them to our easy-to-use site.
> 3.   Create your own descriptions of your home.
>
> Unlike most websites, House Network protects the identity of all our clients by personally talking to all potential buyers and passing on their details to you.
>   We NEVER release your contact details over the telephone and we certainly don't expose your contact details online. We believe that this adds security and saves you time whilst protecting you from unwanted callers.

The House Network website gives detailed tips on how individuals can use the service for marketing the home, after the individuals themselves have done all the "easy work" normally undertaken by estate agents.

For example, this is how a property should be valued:

> Valuing your property is easy by simply comparing similar properties that are for sale in your area. You can find details of these properties in local papers, online or in estate agents' windows. Before deciding to sell your home online you can easily do your own research into the prices of similar homes to yours in your immediate area. The bottom line is that YOU have the autonomy to decide what price to put your home on the market for. Obviously you have to be realistic but at the same time you want the best price possible for your property.

The site also quotes a case study to show how straightforward life is without estate agents:

> When I wanted to sell my home online, I too was worried about the valuation. I discovered that home prices are driven by demand in a

particular town. As long as you base your price on homes either just sold, or up for sale, within your area, you will be pricing your home the same way as an estate agent would. How do I know? After I investigated the market and put a value on my house, I asked a local estate agent for a free quote — and his price was only £1000 different from the value I put on my own home!

House Network allows its clients to submit an unlimited number of photographs of their property for inclusion into the website. They can also purchase floor plans at £40 for properties up to 5000 square feet, created after the vendor sends the firm what it calls "a rough drawing" by fax or e-mail. It promises that within two working days an electronic finished product is then put into the property details on the website and is displayed for an unlimited period.

If an individual doubts their ability to take good photographs, a House Network representative will visit the property and take up to 50 digital shots for another £40; for an additional £50, the same representative will take extra shots and create a virtual tour.

The seller writes his or her own property details. Perhaps, unwisely, House Network promises to give "an unlimited amount of space" and vendors are encouraged to add special sections about local schools, public transport and other area information.

A For Sale board, obviously containing a link to the website but no other contact details, will be put up for £30 although the core advertising medium is the internet. When House Network is contacted by an interested would-be buyer (either as a result of the For Sale board or directly through the website), the seller is contacted and asked whether he or she wants to handle the viewing directly or have it managed by House Network at a charge of £40 for viewing or £50 if the seller wants feedback.

The firm is keen to play up its pro-security approach:

> When we are contacted by the potential buyer, we take down their full personal details, including their name, current address, e-mail, telephone numbers and their current position and status — for example, "first-time buyer with a mortgage agreed". We then contact you and provide you with all of the gathered information on your potential viewer so that you can evaluate them. The decision is then left with you to decide if you want to contact them to arrange a viewing of your house or indeed ask for further information OR you can decide that they are not suitable and choose NOT to arrange the viewing. Our service is fully explained to all callers wishing to view your house.

Once viewings are done and an offer has been accepted, the website advises clients to use a solicitor to handle the conveyancing. But there is an encrypted, secure "after-sales progress page", which sellers and buyers can use to update each party on how the sale is progressing.

This feature encourages far more communication between seller and buyer (albeit through the medium of the website) than in traditional transactions. For example, it urges sellers to advise their buyers when they have notified solicitors and to publish their contact details so all parties can chase progress; likewise, buyers are recommended to inform sellers when they have arranged for a survey, before the sellers are contacted by the surveyor to fix an appointment.

Although there are many extras (the costs of the For Sale board, virtual tours, digital pictures and viewing management, for example) the core service is only £49 including VAT. If you want your property details on not only the House Network website but also the Assertahome and Fish4Homes portals, the total cost increases to £115 plus VAT, which will also entitle the seller to have hard-copy printouts of the property details to distribute through any other route that he or she wishes.

There are plenty of rough edges to the House Network service — it is not always easy to contact them on the telephone, for example, despite their pledge of a 24-hour service centre to handle calls. But it is more the promise (or the threat, as established agents would probably see it) of what such a service could do that is worth taking on board.

Is this the future? If it is, can it be changed by agents themselves using more of the technology espoused by House Network and engaging the client more in the process? As ever, there are no answers — just plenty of questions and challenges.

Case study 9: *www.email4property.co.uk* — technology for the public to contact agents

This is a simple website using simple technology but giving the potential buyer or renter a huge feeling of empowerment. Also, it puts estate agents on the spot if they are slow in replying to their e-mails or, worse still, ignore their inbox completely.

"Finding your next home couldn't be easier," is the slogan espoused by the site. Users simply select any suburb of London from Abbey

Wood to Woolwich, or any city in the UK from Aberdare to York; they then get a list of the estate agents in that area who have registered with the site, and a helpful list of nearby towns and their agents too, in case the individual wants to extend their property search.

Users can then contact any one individual agent or every agent in their area with a description of the property they are looking for.

Prescribed fields on the internet page make the information manageable for the recipient agents, so users have to fill in separate sections to specify who they are, their chosen price range, their status in terms of any existing property that they have to sell, as well as whether they require a mortgage.

Then, they have to specify their usual accommodation requirements, including preferred districts, definite no-go areas, property type and style, the number of bedrooms and reception rooms, plus whether they want facilities nearby, such as schools or bus routes. There is then a "freeform" section for the user to complete, and the site offers a sample e-mail that should be sent to agents:

> Example e-mail: We are a three-person family with a teenage son at college. We both work on the south side of the city and would prefer south if possible. We have a maximum £185,000 budget and will accept a smaller two or three-bed property but it must be in a good area with a garden for the dog. A diner/kitchen is essential but a garage is not. We have a sale agreed with a completed survey and it looks very solid. We need to be out in 10 weeks — send me your most suitable properties without delay.

The power this gives to the user is immense — contacting a large number of estate agents in one e-mail, which cuts out foot-flogging and wasted time. The acid test, which has not been quantified, is whether all the agents contacted are sufficiently motivated and technically enthusiastic to take such enquiries seriously.

It is a taste of how technology can be used to empower the public as well as the estate agent.

Case study 10: Video conferencing — Taylor Woodrow, Bryant Homes and Countryside Properties lead the way

Video conferencing's attractions to the property industry are obvious — lower travel costs, more rapid meetings, and simpler logistics. It has been in existence since the early 1990s and is a communications medium with a £300 million annual turnover, yet property lies far behind many other industries in adopting this technology.

There are exceptions, of course, although mainly among developers rather than estate agents.

Taylor Woodrow has been using the medium since 2001. "Scarcely a day goes by without our using it in my office," says Emma Cordingley, regional director in the Warrington office of Taylor Woodrow Property.

"Typically, the managing director in Staines would want to talk with me about a project and I'd want to discuss it with other parts of the group. One option would be to go down to London but, with flights at £250 a throw, that takes up a lot of money — and more importantly, a lot of time," claims Cordingley.

One of the ways Taylor Woodrow uses video conferencing is as a medium for legal discussions with solicitors working at an external London law practice, who are spared having to travel to Warrington. The result is that Taylor Woodrow saves on travel expenses and solicitors' fees. There is also the mundane but practical advantage of having meetings start at 9am instead of late morning or lunchtime, when London trains and aircraft arrive in the north of England.

Taylow Woodrow Property has video conferencing in its Glasgow, Leeds and London offices as well as in Warrington, and now its subsidiaries TW Construction and Bryant Homes use it too.

Cordingley says the time it saves spreads throughout all levels of the company. "We used to send our secretaries, office juniors and receptionists down to London for health and safety training. The sessions would start at 9am so we'd have to put them up in a hotel. Now we hold three hour teach-ins with them in front of a screen in Warrington — you can see how the savings occur," she claims.

Another supporter of the technology is Countryside Properties.

"It's come into its own with the increased number of mixed-use developments we've been building. We often have to speak to different sections of the organisation in different areas. Without it, you could easily waste a whole day on just one meeting," according to Guy Lambert, Countryside's head of corporate communications.

Countryside has five video conferencing facilities at Leatherhead, Warrington, Brentwood, Bristol and London. At present, it is investigating a new generation of equipment using PC screens and webcams instead of TV screens, as well as pioneering experiments with hand-held PDAs with video conferencing software.

Video conferencing is relatively inexpensive and few large

companies would fail to make a business case justifying it. Our examples here are large developers but you can imagine a multi-branch estate agency spread over a city or region — the weekly meeting or a discussion over business volumes could be much more time-efficient.

A video conferencing system comes as a complete unit, comprising electronics, cabinet, a television set ranging from 20 inches to 36 inches, plus a pan, tilt and zoom camera, which is operated by a handset similar to a TV remote control unit. The total cost would be between £3000 and £7000 for two units — one for each end of the discussion — plus much cheaper proportionate costs for subsequent units.

The arrival of broadband telephone lines makes the capacity for video conferencing much greater and allows it to reproduce high-definition pictures, giving greater scope for presentations. Broadband also allowed PCs to be used instead of televisions.

Lambert says video conferencing cannot replace every meeting — personnel issues and site visits have to be done in person — but he believes the industry could do much more.

"I'm constantly amazed at how few property companies use video conferencing," comments Taylor Woodrow's Emma Cordingley. "As an industry, we're light years behind some others."

Learning the lessons

Those 10 pioneers are not all successful and are not all role models, but at least have been innovative and pushed the boundaries, and may give a taste of what can be done.

The next chapter gives details of the rewards to be won by property professionals who are the bravest at adopting new equipment and giving themselves a high-tech makeover.

A larger share of an enlarged business, a better reputation in the eyes of the public, and, perhaps, even higher fees. It is yours for the taking.

E-everything — What Could Be Achieved If Only Agents Embraced New Technology

The biggest prize for estate agents is still to be won.

Whatever the future holds, using good-quality technology will always encourage more buyers or sellers to use estate agents simply because it looks efficient and effective. But there is much more business that could be won and, unlikely as it may seem, it is the government's determination to modernise the property transaction process that creates the best opportunities for estate agents to expand their business, reduce their overheads and embrace technology.

Under the government's e-conveyancing and e-selling initiatives, almost all of the traditional house-selling processes are to be renegotiated and the best-placed estate agents are already seeking a much larger slice of the action. Put bluntly, the technology exists for estate agents to undertake at least some of the conveyancing, searches, valuations and other activities central to house selling that have, hitherto, been left to other parties.

But agents must move fast. Many other professionals are getting in on the act and in this chapter we will see how some conveyancing solicitors, historic search companies, new entrepreneurs and even public bodies, such as local authorities, have shed their staid reputations by using new technology to join the e-revolution.

The question is ... why haven't estate agents done the same?

E-business, e-government, e-future

The general public's enthusiasm for new technology is much greater than many have suggested. In the second quarter of 2004, 52% of households in the UK (about 12.8 million) had the ability to access the internet from home, compared with just 9% (about 2.2 million) in the same quarter of 1998.

In July 2004, 58% of adults in Great Britain had used the internet at least once in the three months prior to a survey conducted by the Office of National Statistics (ONS). The most common use of it was to send e-mail (85% of respondents used the net for this reason) or to find out information about shopping (82%). The most frequent place of access was the person's own home (82%) although 42% of people had access at work as well, or instead of, the home, and used the web for personal purposes as well as business ones.

In a separate ONS survey in July 2004, about a third of UK adults confessed they had never used the internet. Of these, 48% stated they did not want to use it, or had no need to do so, and had no interest in the internet; 37% had no internet connection while 32% felt they lacked the knowledge, confidence or opportunity to learn how to use it.

Perhaps unsurprisingly, the proportion of adults that had used the internet in the three months prior to the ONS survey was highest in London and south-east England (both 64%) and lowest in the north east of England (43%).

The highest increase in internet access between 2001 and 2004 was in Scotland (where usage increased by 16%), followed by the north west of England, Yorkshire and the Humber (all of which rose by 12%).

Men were more likely than women to use the internet but the gap between men and women closed over the two years to 2004. Among women, the proportion increased by 12% to 54%; for men the total rose 7% to 61%.

Internet use also varied by age. Unsurprisingly, it was highest in the 16–24 age group (83%) and lowest in the 65 and older age group (where only 15% of the population used the internet). In the two years to 2004, the greatest increase was seen among adults aged between 45 and 54, increasing by 13% to 63%. The lowest increase was among those aged 65 and over — up only by 6%, although this effectively doubled the usage to 15%.

The ONS found that half of all the adults who had used the internet in the three-month period under assessment had bought or ordered tickets, goods or services. A higher proportion of men (53%) than women (46%) had used it for this reason.

Adults aged between 25 and 44 were the age group most likely to purchase goods over the internet (55% said they had done so) while the least likely age group was 65 and over.

In a separate survey by the British Retail Consortium — completed in early 2004 but looking at technology usage by the public in the build up to Christmas 2003 — it was revealed that internet usage reached a new all-time high in November 2003.

No fewer than 44% more people used the internet for shopping during that month than they did in the same month the previous year, even though the total volume of retail sales in the same period rose by only 3.6%. This means that, while the simple volume of shopping increased a little, the real change was that people were willing to abandon the dubious distinction of physical shopping in the Christmas crowds, and instead used their PCs and laptops.

In November 2003, Britain's 16 million online shoppers spent £1.175 billion online — that accounted for 7% of all retail sales in the UK. That figure is bound to have increased significantly in 2004.

Only one retailing survey was conducted in 2004, again by the ONS, which suggested that in the three months to August some 19% of UK internet users had bought items worth £500 or more on the web. So much for people allegedly being wary of a lack of security on the internet.

There is relatively little thorough research done into how this revolution in internet shopping has played out in the world of property, although, once again, the National Association of Realtors (NAR) in the US has tried.

The California Association of Realtors or CAR (the west coast division of the NAR) conducted large-scale research in late 2004. By contrasting its results with those of a similar survey conducted in late 2000, it produces one clear message to all estate agents in the UK where internet usage follows almost identically the US pattern, but with a three to five-year time-lag.

The message is that old-fashioned house buyers who do not go online to assist in their property search are in a minority.

The CAR found that 56% of all buyers used the internet when buying a home, exactly double the 28% finding of 2000. As with previous surveys, CAR found that internet buyers did not use it to save time as such — indeed, they spent almost twice the number of hours researching how, where and what to purchase than their "traditional" counterparts did. But that means that buyers who went online became much better informed by the time they short-listed properties and physically visited them — and by the time they made contact with vendors' estate agents.

Other highlights of the survey included:

- internet buyers spent an average of 5.9 weeks considering the purchase of a home before contacting a realtor compared with 2.1 weeks for traditional buyers
- internet buyers spent an average of 4.8 weeks investigating homes and neighbourhoods prior to contacting a realtor compared with 1.6 weeks for traditionals
- once they started looking "in person" internet buyers were much more focused, spending just 1.9 weeks visiting properties before making an offer compared with 7.1 weeks for traditionals
- the typical internet buyer visited fewer homes accompanied by a realtor than did traditional buyers (6.1 compared with 15.4)
- internet buyers were three times more likely to be first-time buyers (23% against 7%), bought more expensive homes (an average of $168,540 compared with $142,470) and were better educated.

It would be naïve to think the same scale of internet usage will not happen in the UK property industry, particularly as this country is set to become a world leader in the roll-out of high-speed internet services to the population at large, according to a report from the Organisation for Economic Co-operation and Development (OECD).

ADSL, a high-quality telephone line that enables the carrying of rapid internet signals, is available on 95% of UK telephone lines, a figure matched only by Finland and bettered only by Belgium, Denmark and Switzerland. BT expects its broadband service to reach 99% coverage by the end of 2005 and the OECD's report claims that no other country will equal this.

The target looks attainable and the public are responding. More than 600,000 UK homes and businesses were linked to broadband in the three months to October 2004, taking the total to 3.3 million. In late 2004, the telecom regulator Ofcom revealed that Britain had 7.5 broadband connections per 100 people, ahead of Germany on 6.7 — despite the fact that most private broadband suppliers in the UK were significantly more expensive than their mainland European counterparts.

A few years ago, it seemed unlikely that the UK would reach this position. In 2000, ADSL was available to over just 50% of the population compared with 60% for Germany and 69% for Canada but demand by business users and a growing band of teleworkers has pushed this country to the fore.

The government has responded. The New Labour administrations elected since 1997 made successive clear commitments to make vast swathes of public sector activities accessible on the internet.

As a result, you can go online to undertake the following public service activities that, as recently as 1998, could be done only through telephone calls, personal visits or letters:

You can ... contact the police, find and recruit a legal advisor, report a consumer problem, find a job or check your entitlement to redundancy payments, get on the electoral roll and register for a postal or proxy vote, nominate a person for an honour, register birth or death or marriage, apply for lottery funding, review available health services in your area or overseas, check health performance tables and waiting lists, claim financial assistance for health costs, make a complaint against the NHS, apply for, or renew, a passport, check the 2001 census or trace a family history back to 1901, research war casualties since 1901, buy a fishing rod licence, check tidal information in the UK and abroad, get UK weather forecasts, buy or renew a TV licence, buy coins, register as a voluntary worker, search for authorised childcare facilities in your area, register to become a school governor, find a list of local schools and their Ofsted reports and league table performances, find and register for adult learning courses, take a mock theory driving test, book a real driving test, pay the London congestion charge, find out about UK-wide traffic information, check whether your vehicle has now, or has ever, been recalled, buy a personalised number plate, apply to register a new or second-hand car, seek exemption from an MOT, apply for a Heavy Goods Vehicle licence, buy an export licence, get help covering legal costs, seek online advice from a solicitor, apply for British nationality, find your local MP, complain against your local authority or pension provider or utility company, compare rival council performances, apply for a disabled person's railcard, apply to become a special constable or to join the police force, or apply to an industrial tribunal.

You can also pay your national insurance and income tax online, safe in the knowledge that what you pay will reach the government more quickly. Well, there is a downside to everything.

But government involvement in the residential property industry is at the forefront of its e-targets. One of the largest projects is to modernise planning permissions: it is now notoriously slow, at least partly because of the high levels on consultation and town hall paperwork involved in every application.

During 2003, there were around 620,000 planning applications in England, of which about 80% were submitted from the business and voluntary sectors, including planning agents and the construction industry. It is easy to see that if five or more interested parties were also involved in commenting on each planning application (probably a conservative estimate) then this would create over three million transactions.

To speed up planning processes and reduce paperwork, the government has set ambitious targets for applications and allied processes to be conducted online. For example:

- by the end of 2005, 10% of all pre-application enquiries will be made online, rising to 90% by the end of 2008
- by the end of 2005, 10% of all planning applications will be submitted through electronic channels, rising to 60% by the end of 2008 and 90% by the end of 2011, with incentives for users (likely to be a £10 discount compared with regular fees)
- for planning appeals against decisions, 10% will be submitted electronically by late 2005, reaching 95% by the end of 2008 and 98% by 2011
- 10% of planning enforcement cases will be dealt with electronically by late 2005, rising to 65% by late 2008 and 90% by 2011
- 20% of all building regulations and controls enquiries will be electronic by the end of 2008, rising to 50% by the end of 2011
- enquiries to the Planning Consultation Service offered by town halls will also change, hitting 10% by the end of 2005, 90% by the end of 2008 and 95% by late 2011.

Although changes of this kind are theoretically making public service activities more accountable and efficient, fuelled by more and easier public participation, they have, perhaps, failed to make a huge impact in everyone's lives so far.

The same will not be said about the changes coming up during 2006 and 2007 in the residential buying and selling business. In that instance, everyone in the industry will have their working practices and lives transformed by the largest revolution since the start of estate agency.

The more people adopt and train for the technology that is at the heart of the revolution, the bigger part they can play in it.

Home Information Packs and the revolution in estate agency

Agents could do almost all of the organisation of the packs if they wanted to and had the right systems. It needs positive thinking and a really small amount of investment. But don't hold your breath while they decide whether they'll do it.

Those words, from a leading estate agency software supplier who chose not to be named, highlights both the scale of the prize that agents could capture and also the cultural reluctance of the residential industry to embark on something new.

Whether the professional agents or their umbrella groups want the Home Information Packs (HIPs) or agree with the concept of preparing "up front" information, is irrelevant: the fact is the packs are on their way and there is plenty of additional business that estate agents can undertake if they wish, complete with additional fees to be charged.

The government's view is that the current process of buying and selling a house is "shambolic", in the words of housing minister Keith Hill. He says it serves no one well, including the seller.

"We are clear that making key information available right at the start of the process makes home-buying and selling easier, more transparent and more successful," he says. HIPs "will enable buyers and sellers to exercise real choice at a stage when they are not yet committed — a fundamental principle of all consumer legislation."

The Office of the Deputy Prime Minister (ODPM), which is supervising the introduction of the HIPs, says every new build and second-hand property will have a pack before being put on the market. Each pack should contain:

- *Terms of sale*: that is, whether the property is offered freehold, leasehold or commonhold. There must be confirmation that it is being sold with vacant possession on completion, the exact agreed purchase price and an agreed completion date, plus a list of fixtures and fittings. The form of the contract would be flexible but the government is currently considering whether to include a draft contract in with the pack.

  There are already some specialist software companies preparing online pro formas of this part of the pack, which could easily be completed by an estate agent.
- *Evidence of title*: this includes official copies (which must not be more than three months old at the point of initial marketing) of the property register, proprietor register and charges register. There should also be a title plan and copies of any documents and plans referred to in the registers. Where the title is not registered,

the pack should contain copies of a "Certificate of Result of Search of the Index Map" that must also be under three months old at the point of initial marketing. The title documents should also include deeds and other paperwork held by or under the control of the seller, providing evidence of the title on which the seller intends to rely should there be a legal challenge.

An estate agent in alliance with a conveyancing solicitor would be able to ensure this was undertaken for a client.

* *Searches*: the pack will contain replies to local land charge search enquiries and to other standard additional search enquiries concerning the property. These enquiries should be set out in a standard form or forms, and the replies should be not more than three months old when marketing of the property commences.

This is often cited as the area where estate agents could not get involved — but as the latter part of this chapter shows, an agent with online capabilities, training and subscriptions to search services could play a major part and greatly boost earnings.

* *Planning permissions and building regulations*: the pack should contain copies of any planning permissions (including any conditions) granted under the Town and Country Planning Acts relating to the property, plus details of listed building consents or conservation area consents. For new or nearly-new homes, any agreements under section 106 of the Town and Country Planning Act 1990 relating to the property or a development of which the property forms a part must be included, as should directions under Article 4 of the Town and Country Planning (General Permitted Development) Order 1995 affecting the property; this part of the pack should also include certificates given under the Building Act 1984 relating to the construction or alteration of the property.

Slowly but surely local authorities are putting this information online and available to conveyancers — this could mean estate agents could do this part of the pack preparation, where previously they would have deferred to traditional conveyancing solicitors.

* *Seller's property information form*: the pack will include a prescribed form for use by sellers providing answers to standard enquiries made on behalf of prospective buyers. This form should be concerned with providing information on matters likely to be important to potential buyers but which may not be contained in the evidence of title — in general terms, this is likely to be information on boundaries, disputes, notices, services, sharing of facilities or surfaces with neighbours, arrangements and rights,

occupiers, restrictions, planning, fixtures, expenses and any other matters about which the seller would normally be expected to have information and which are important to a buyer.

Currently, the favoured document to satisfy this is one drawn up by the Law Society but it is often handled by estate agents.

- *Warranties and guarantees*: the pack must include any current National House Building Council or Zurich Municipal warranty or equivalent (usually lasting for 10 years from the completion of a modern property, assuming the developer was a member of one of these warranty schemes). This part of the pack should also include any contract or insurance policy relating to the fabric of the property, which would routinely transfer to a new owner.

- *Home Condition Report (HCR)*: this was the most controversial part of the pack although most interested parties — lenders, surveyors and even some estate agents — came to accept its existence by late 2004. The HCR is defined as an objective report on the condition of the property. This should be prepared by a Home Inspector (HI) qualifying under a certification scheme approved by the Secretary of State. The report should include an energy efficiency assessment compliant with EU Directive 2002/91/EC (Energy Performance of Buildings Directive).

  The cause of concern in the years leading up to the pack becoming law was that many industry players wondered how the HCRs would complement — or conflict with — the traditional lender's surveys. Would, for example, a lender preparing to offer a mortgage of between £50,000 and perhaps £3 million allow a seller to commission a report which would then be the basis for a home loan? If the answer was Yes, what would happen to the "old" surveyors? If the answer was No, would this not lead to substantial duplication of the HCR and the long-standing lender's survey?

  Whatever the rights and wrongs of the principle, the process is now agreed. A recruitment campaign is under way to ensure more than 7500 new inspectors are in place to prepare the reports, which will be written online.

For leasehold properties, in addition to all those provisions above, the pack must contain:

- *a copy of the lease*, together with any head-lease(s) or sub-lease(s) of part(s) of the property
- *copies of the most recent service charge accounts* provided in accordance

with any rights to receive information under the Landlord and Tenant Act 1985 (as amended) and receipts and ground rent receipts

- *details of current and planned future works to be funded by service charges* where a formal notice has been issued by the landlord under the Landlord and Tenant Act 1985
- *copies of the insurance policy* covering the building and receipts for premiums
- *a copy of the current regulations* made by the landlord or management company
- *copies of the memorandum and articles of the management company* where this company is deemed to be controlled by the leaseholders.

There are plenty of spin-off opportunities for estate agents resulting from the packs. For example, some could:

- Simply act as "collators" of the pack — ordering, chasing and then collating the information that would come from a variety of third party sources (probably online).
- Actively gather some of the information themselves — for example, they could become HIs so as to write the HCRs, and/or they could employ a conveyancing solicitor in their office to form more of a one-stop shop for clients, and/or they could obtain some of the search information themselves by using technology to obtain the information online.

Agents who want to collate the pack need to act quickly if they are to stop the corporates getting a foothold — and we do not mean merely the corporate estate agents. Large organisations, hitherto on the fringes of estate agency are trying to take a huge slice of the pack preparation business.

For example, property internet portal Rightmove says its aim is to be the largest provider of HIPs by the time the HIPs are compulsory for all homes on the market in early 2007.

The company says it is preparing to provide at least 500,000 packs a year from the end of 2006. This figure is thought to be enough to satisfy over a third of the total number of packs required annually by home sellers in England and Wales, based on the average number of 1.4 million property transactions a year over the five years to 2003.

Rightmove has been owned since early 2000 by many of the estate agency corporates — Countrywide Assured, Connells, the Halifax and Royal & Sun Alliance — and it has some 6000 agencies that subscribe

Section D - Exterior condition

SAVA

52 High Street, Hemel Hempstead, Hertfordshire, HP1 3EG

Home | Previous | Next | Close | Navigator | Getting Started

The Home Inspector has not been able to inspect the following parts of the exterior of the property for the reasons stated here

Chimney stacks
Condition rating — 2

Justification for rating or comments — The chimney stack is generally in sound condition but the condition of the jointing of the bricks in the chimney stack (known as pointing) needs minor repair. Failure to undertake this work may affect the

Roof coverings
Condition rating — 1

Justification for rating or comments — No repair is presently required. Normal maintenance must be undertaken.

Rain water pipes & gutters
Condition rating — 3

Justification for rating or comments — Urgent attention is recommended to prevent damage to the main building

Main walls
Condition rating — 1

Justification for rating or comments — The external walls are of solid construction using brick.

Terms and conditions · Support · Privacy

Home Information Packs are on their way, irrespective of the controversy they caused before they became law. The Home Condition Report, a critical element of the HIPs, will be completed online by trained inspectors. Companies which market "pack preparation services" to agents and individual sellers say the full HIP may well be online too.
Source: *SAVA*

and advertise on the Rightmove website, including major high street chains such as Your Move and Spicerhaart. They cannot be forced to use Rightmove, but there may be incentives for them to do so.

Ed Williams, group managing director, says the initiative is logical: "Whatever the pros and cons, there is no question that every estate agent will be legally required to have a Home Information Pack available before starting to market any property. We have taken the decision to help solve what looks like being a massive headache for every estate agent."

Williams says Rightmove was asked in 2004 how new technology, including use of the internet, could "reduce the pain and cost" associated with having to prepare large numbers of packs.

"Having helped estate agents maximise the potential of the internet it seems a very natural step to support our customers and the estate agency industry in this challenge too," according to the publicity-savvy Williams. Even in the days before the HIP legislation received royal assent he boasted that the Halifax estate agency, Bairstow Eves and Connells were definite pack customers.

Jane Pridgeon, managing director of Halifax Estate Agents, says: "One of the big attractions to us is that Rightmove has a proven track record in producing cost-effective technical solutions that work for both the smallest and the very largest estate agency businesses."

First Title, the UK's best-known provider of title insurance and remortgage services, is formally a financial organisation and comes under the regulation of the Financial Services Agency, yet it too is aiming to be a major HIP provider by drawing on the expertise of its UK parent company First American.

"We all appreciate that the US property market is very different to the UK's [but] our parent is unique in its ability to gather data and provide it to others in a fast, cost-effective manner. In one subsidiary alone it manages 600 million title documents," says a First Title spokesman.

It may be that many estate agents will be happy to allow packs to be prepared by third parties, such as Rightmove, which was reported to be creating an infrastructure costing over £10 million to handle its diverse business interests, or by First Title, which says it has spent an unspecified seven-figure sum on investment.

But many agents may also want to prepare the packs themselves — after all, if non-agents like those mentioned above are confident of making a profit by preparing HIPs, there must be a genuine opportunity for agents too.

Those who do want to complete the full pack themselves, or at least all of the pack except for those elements that need extensive and strict legal involvement, will have to undergo a rigorous training programme.

The most extensive training will be to become HIs qualified to undertake and write the HCRs. HCRs will be 20-page online documents describing the condition of all the parts of a home.

Current surveyors are not automatically entitled to act as HIs. They and all other applicants must train, mostly by "attending" online training seminars, for between six months and two years to obtain a new Home Inspector Diploma qualification. Applicants will be chosen by independent assessment centres being nominated by the government.

The ODPM says HIs are expected to come from the construction, property and surveying back-grounds as well as attracting college and university leavers with surveying qualifications. But the ODPM expects some estate agents to apply in a bid to boost their involvement in the overall property-transaction process.

The Awarding Body for the Built Environment, an academic standards body that oversees many surveying and construction industry training, has decreed that the national occupational standards for the HI role will be the equivalent of NVQ Level 4 (HND or basic, non-honours, degree level). Inspectors will obtain a licence by entering and passing a final competency test that will be considerably more difficult than current surveyor training. It will include areas currently missing from traditional residential surveys, such as energy efficiency, and health and safety. There is also a module of training in, and marks awarded for, the ability to write a survey report in plain English with fewer caveats and less technical jargon — so that will be a real break from tradition.

The government hopes HCRs will replace the traditional surveys that occur in 75% of residential transactions to satisfy the concerns of mortgage lenders (in the other 25% of transactions there is no mortgage, so it is discretionary for the buyer to instruct a surveyor — most of these buyers do not bother).

"HCRs will be far more rigorous than old-fashioned surveys. Often the surveyors do not even go into the house and they just check that it looks okay. HCRs will be much tougher," according to a spokesman for the ODPM. The office estimates that for a post-1960, two- or three-bedroom home with internal floor space of around 100 square metres, the inspection alone will take upwards of 1 hour and 45 minutes, and the report writing will take a predicted further three hours.

HCRs will be filed electronically to a national database so future sales of the same property can call on previous reports and judge how its condition has changed over time.

The Halifax, and some smaller mortgage companies, said as early as summer 2004 that they would accept HCRs as the basis for deciding whether to lend to buyers, without the need for a duplicating traditional survey of their own. The government says other lenders will follow suit before the HIPs become mandatory in 2007 — a move that is critical to avoid some lenders continuing with existing surveys, thus effectively duplicating the inspection process and undermining the HIPs.

The first draft of what an HCR will look like has been prepared by the ODPM and it consists of:

- **Section A** — the terms and conditions, definitions and explanations for the vendor, on whose behalf the HCR is written.
- **Section B** — general information about the property summarising its size, age and overall condition.

- **Section C** — called "other matters", looks at health and safety issues, such as quality of repair work through staircase condition insides and the presence of ponds or lakes in the grounds, for example.
- **Section D** — the exterior condition, from chimney stacks to windows and doors, from cladding to roof coverings.
- **Section E** — the interior condition, with room-by-room descriptions from the roof space to the cellar, and issues from dampness to bathroom fittings.
- **Section F** — public and private utility services, including electricity, gas, water, heating and drainage.
- **Section G** — the grounds, boundary walls and outbuildings.
- **Section H** — a new and detailed energy efficiency report.

The exact length of training required to obtain an ABBE Home Inspector Diploma and then a licence to practise as an HI will vary according to each applicant's experience but all will have to take place through an official Home Inspector assessment centre. Much of the training will be part-time (after all, most applicants will have current jobs to continue).

The first assessment centre, run by Sava, has set out a five-point training programme which is typical of those that will take place until the introduction of HIPs. Thereafter, will take place as a running programme as active inspectors retire or leave the profession. Notice how much emphasis is placed on using new technology:

- *Stage 1: An experience scan* — this assessment categorises candidates as "existing practitioners" or "new entrants" and ensures they have sufficient knowledge of residential properties to even embark on training. It is expected that most chartered surveyors with long-standing experience of home surveys and similar professionals will be accepted without difficulty. People with no experience in residential property or a related industry are likely to need further education before they can be admitted to formal training for the Home Inspector Diploma.
- *Stage 2: Self-tests* — all candidates will have access to software to test themselves on every element of the qualification. This provides an early taster of the rigorous end test (set up by the supervising Awarding Body for the Built Environment) and will identify where further learning might be needed. Any such top up tuition will be online.

- *Stage 3: Building an e-portfolio* — candidates must create and manage a portfolio of their work (which is defined as a body of evidence to demonstrate their competence, knowledge and skills) using online systems. Into this e-portfolio will go details of applicants' academic qualifications, practical work experience, sample surveys and any new training that helps to demonstrate their competence. The evidence is uploaded via a candidate's PC at home or in the office; the assessor gets access to the information instantaneously and any feedback from the assessor is also electronically relayed back to the candidate. Because the assessment centre is online it is always open, meaning that both the candidate and assessor can work on the portfolio at any time.
- *Stage 4: Using the Home Condition Report software* — as part of the assessment process, candidates must produce 10 residential property reports of which at least three must be in the new HCR format, discussed above. The new software allows HIs to produce an approved HCR, including the detailed energy report, for each home surveyed. These must also include recommendations on how the property energy performance can be improved.
- *Stage 5: The 90-minute end test* — after a candidate's e-portfolio is complete, the end test is the last hurdle. It will include a mixture of multiple-choice questions and case studies based on properties ranging from Victorian semis to modern homes.

While the HCR is by far the most technical element of the HIP contents, it is just one of the sections that estate agents may want to consider providing themselves.

For the remaining sections, there is a wealth of online services available to help in preparation.

21st century electronic infrastructure
National Land Information Service

This is a joint initiative between central and local government that provides an online one-stop shop that can deliver land and property-related information to a PC in a single package, even though the information is actually owned and held by many different organisations.

Initially, National Land Information Service (NLIS) is concentrating on providing a service to the conveyancing community, in order to speed up and simplify the process of buying or selling property (mainly but not wholly residential) but it hopes that increasing numbers of estate agents will also subscribe. By 2008, NLIS aims to allow all searches to be carried out via the internet.

Among the information suppliers it calls upon is the National Land and Property Gazetteer (NLPG), a definitive index of land and property (NLPG — section c of this chapter). NLIS also offers online access to large-scale Ordnance Survey digital mapping, which also gives the ability to define the property boundary on a map, which can then be submitted with search requests where required. So if you know the boundary of a property and can set it out online, it ensures any consequent search is about exactly the right footprint.

Any conveyancer using NLIS can also retrieve online ownership information from the Electronic Land Registry as well as from all local authorities that are pertinent to the property. In addition, there is the facility to search for other information — even quite unusual data is now online, such as mining histories from the Coal Authority.

In the future, the bringing together of this wealth of property information is likely to be packaged in a variety of ways to provide services that are tailored for different property professionals, be they surveyors, estate agents, mortgage lenders, developers or insurers. NLIS will be that flexible.

As is commonly the case today, much of the traditional search information will continue to be held by public authorities. Indeed, NLIS itself is government funded, but the way which any estate agent or other user would access this information would be by entering into a contract with one of the three privately owned "franchise holders" which control the access channels.

By buying into these channels, conveyancers can request and receive their online information through pay-as-you-use or long-term contract arrangements.

One channel company is Property Services Ltd, a joint venture between management consulancy KPMG, Countrywide Assured estate agents and a number of private equity investors and venture capitalists. A second licence holder is TOL, a wholly owned subsidiary of MacDonald Dettwiler Ltd, a global technology giant. The final licence holder is Searchflow, part of The Conveyancing Channel, a joint venture between the Property Search Agency (one of the UK's leading conveyancing search consultancies) and ESRI Products Ltd, a supplier of local authority search systems.

Each channel has a noble but complicated objective — that is, to allow the most bureaucratic elements of house buying, such as council and public utility searches, to be carried out quickly on a computer or laptop instead of through the post or via long telephone calls.

"The key advantage we have is certainty. A conveyancer can see a property on a map, enter its address and be 100% sure that we're getting all the data associated with it. A skilled user can enter this information within as little as one minute and certainly in no more than 10 minutes, and then await the results," says Searchflow's chief executive officer Mark Riddick.

His system is typical of the others and has six straightforward elements.

First, you log on to *www.searchflow.co.uk* and enter the username and password, which you set up on your registration with the service.

Second, you enter a client reference number (either an old one to chase up an existing search, or you create a new one for a new client).

Third, you enter the address of the residential property to be searched. NLIS automatically validates the address to avoid you accidentally searching for a wrongly entered address.

Fourth, you mark up the extent of the search. That is done by using a map and simple instructions, on which you indicate precisely the footprint of the property on which the search will be conducted.

Fifth, you select the range of searches — should they include Ordnance Survey and other data on NLIS or do you need to seek more specialist sources such as the Coal Authority as well as the more obvious local councils and private utility companies? Finally, after submitting this request, Searchflow lists your requirements and the sources, along with the cost of the total search, for you to check before confirming and commencing the exercise.

Searchflow sees a big savings opportunity for large legal practices that undertake hundreds of searches. "Sometimes it costs £50 to issue a cheque for a £30 water search. That's madness. Eventually, Searchflow will allow multiple-order billing between the user and all the data providers and make these transactions much simpler," promises Riddick.

"The legal profession is slowly getting to grips with technology, in large companies in particular. But if you look in the window of some offices you still don't even see a computer. That's scary but we have to convince them that working online offers them a simpler and quicker process that benefits them and their customers."

There is a further difficulty. Not all local authorities have full-scale online search facilities, so companies like Searchflow and the other

NLIS access channels can be frustrated as they try to provide users with the quickest possible service.

One user of Searchflow is David Pettingale, a partner in the large Birmingham conveyancing solicitors' practice Wragge and Co.

"Remember that the service is a shadow of what it will be. For example, some local authorities respond to a search query in two or three days because they've invested in technology to complete the process quickly. But many are still using the old manual systems and are taking two to three weeks, even if they then deliver the results to you through an online interface that looks more modern than the clunky interface behind it," he says.

"Eventually, that will change and the service will be genuinely quicker to the paying customer."

The genesis of NLIS can be traced back to the late 1980s and an initiative by the Royal Institution of Chartered Surveyors. By the mid-1990s, it became a central plank in the Labour Party manifesto, subsidiary to (yet essential for) making the house-buying process much simpler.

In 1998, a pilot system was introduced in Bristol and in December of that year it received full cash funding from the government. The "Modernising Government" White Paper, issued in 1999, set a target that by 2002 "25% of dealings with government should be capable of being done by the public electronically" and furthermore "...we propose that 50% of dealings should be capable of electronic delivery by 2002 and 100% by 2008".

Despite the popular belief that conveyancing solicitors are among the least technically savvy players in the residential property industry, figures show they have been relatively quick to take up NLIS. By summer 2004, over three million searches had been made online by lawyers on behalf of homebuyers in England and Wales alone.

In the year to August 2004, NLIS processed over two million searches compared with the one million that had been processed during the combined two previous years, and, according to the service, it was handling upwards of 30% of all residential transactions.

"This is one of the biggest and best examples of e-government in action. NLIS is not only achieving dramatic improvements in the quality and responsiveness of services but also joining up the private and public sectors. It is improving services for the public and helping local authorities to improve their efficiency. It is also the reason why NLIS is now one of the fastest-growing internet legal services ever to be embraced by lawyers, with over 6000 practices ordering searches online," according to the enthusiastic James Thornton, e-government director at NLIS.

There is little doubt that it is a time saver and one that has kept up with legal changes. For example, coinciding with the Commonhold and Leasehold Reform Act 2002 that came into effect in late September 2004, all three NLIS channels launched a new search service for commonhold and leasehold titles held by the Land Registry.

Alison Walker, president of the Local Land Charge Institute, sums up the way in which this part of property transacting has changed.

"NLIS has bought local land charges into the 21st century of e-government. We handle over a third of all of our searches via the NLIS service, drastically cutting out two days of turn-round time. I feel that NLIS is the way forward for councils to compete in a rapidly changing world, providing an even more efficient, professional service to solicitors throughout the country," she says.

As the NLIS service grows further to that 2008 target the government envisages that companies using Searchflow and the other NLIS channels can generate additional fee income from value-added services. These could take the form of a change of address service, advertising, or a commission from the sale of related products and services, such as removals or loans.

Currently, NLIS is aimed mainly at conveyancing solicitors — but could estate agents use it if they could be proven and skilled supporters of the technology?

Electronic Land Registry

The government is convinced that long-standing Land Registry working practices make conveyancing slower, more complicated and more expensive than it should be. It points to the three stages of the current process, based on an 80-year-old law:

- First, a transfer document has to be created as described in Land Registration Rules legislation dating to 1925.
- Second, the document must be lodged at the relevant district Land Registry office.
- Third, the Land Registry processes the document — taking no fewer than 13 working days on average (in reality, much of this time involves the document sitting in a pending tray as the physical check takes just a few moments).

The delay in this last stage is known in the industry as the "registration gap" and can create problems if a separate search is also conducted within that time for, say, a buyer of nearby land or property.

In 2001, the government issued a 200-page consultation document (online, of course) outlining the advantages that would accrue from modernising the conveyancing process:

- Transactions would be quickly and substantially cheaper for buyers and sellers and the risk of error significantly reduced.
- Problems caused by the registration gap would be reduced.
- Large-scale and expensive paper storage would no longer be needed.

Most of the industry is throwing its weight behind the government's plans although there remains some uncertainty over how long the objective will take to fulfil. "We know the legal and conveyancing professions need to modernise and we're trying to educate our members along these lines," says a spokeswoman for the Law Society.

"We contributed to discussions begun by the government on e-conveyancing and we've started the ball rolling with our members by issuing new guidelines over e-mail correspondence and trying to alert people that there will be these large scale changes over the next decade or so," she says.

The government says three key issues must be overcome before e-conveyancing becomes the definitive land and property registration system:

- *Security* — it must be guaranteed that the only people who can create electronic registration documents are recognised by the Land Registry as bona fide users; this is a hard nut to crack because the government remains determined that individual citizens should be able to do their own conveyancing, as now, but also is conscious that security of information is important.
- *Accuracy* — the government suggests there should be Land Registry training when the system is introduced for anyone who wishes to join the conveyancing industry after that point, and this could apply to current estate agents if they apply.
- *Technology failure* — systems supporting such a large database need to be robust and not subject to the problems that have beset so many other areas of government modernisation, ranging from

passport application processing to driving licence renewals and even air traffic control.

Some conveyancers, large and small, have used electronic systems for some years. For example, the Nationwide building society and the Land Registry first introduced a pilot system in 1999.

It allows property registration certificates and mortgage deeds to be held in electronic format only by the Land Registry, with no requirement for the Nationwide to keep hard copies. If the Nationwide needs a copy for further change of ownership or for client queries, the Registry sends an electronic version.

In turn, solicitors working on sales or purchases funded by the Nationwide receive electronic copies of Land Registry documents, instead of the originals or paper copies.

At the other end of the size scale, the Nottinghamshire-based law firm Fidler and Pepper has given conveyancing quotations online since 1995 and since 1998 has allowed internet-based "case tracking" — a process whereby clients buying and selling property can monitor the progress of their conveyancing by getting into a secure part of the website.

"It's attracted business from London and other parts of the country. Conveyancing essentially doesn't require legal advisers to be nearby. You could have a conveyancer 100 yards away but if they're always on the telephone and you cannot get through, they may as well be 10,000 miles from you," says a spokesman.

"Case tracking means that we have 20-minute updates to a secure site which can be accessed only by clients buying or selling — literally tracking their case. As a result of this, our solicitors have far fewer telephone calls and those that do come through concern very specific queries. We don't get calls from people just wanting the general state of play," he says.

"The only problem is we're still constrained by the slowness of the conveyancing process — at least for now."

They are not alone. North of England property website *www.premierlocation.co.uk* has teamed up with Dickenson Dees, a major law firm, to provide immediate online conveyancing quotes. A real-time case-tracking system will also mean that cases can be tracked at any time online and customers can save time by using the service by e-mail instead of telephone and post. They will also have the benefit of an automatic text messaging service on exchange of contracts and completion.

If you are an enterprising estate agent, perhaps working closely with a conveyancing solicitor, could you use new technology to provide this kind of service? If so, might it increase your share of business?

National Land and Property Gazetteer

The National Land and Property Gazetteer (NLPG) is the first, definitive, national address list that provides unique identification of properties and conforms to BS 7666 (the British standard that specifies a format for holding details on every property and street and the only one acceptable in conveyancing law). It allows organisations across the public and private sector to link their data to the same high-quality source of references that define locations and addresses.

NLPG is a joint venture between the Local Government Information House (a public sector body promoting e-government at local authority level) and a private firm, Intelligent Addressing Ltd. It is being developed in two stages.

The first is the creation of a national reference list that matches data held by the Land Registry, the Valuation Office Agency and other bodies, against addresses held by local authorities. This is to ensure there is no longer confusion over the exact address of a property, its boundary footprint, its name or number, it planning status, or its ownership.

The second stage is to create a standardised local land and property gazetteer for each local authority that will continuously feed new information to ensure NLPG is always contemporary.

National Street Gazetteer

The National Street Gazetteer (the NSG) is, in many ways, a sister service to the NLPG and is designed to be the most authoritative and unambiguous referencing system that can identify any street, road or public space in Great Britain.

It conforms to BS 7666 and provides a complete list of all streets with their names and other useful data created, updated and maintained by local highway authorities — normally county councils. It is a key tool in speeding up searches and, because it is updated by those authorities, it can help identify where there have been plans for major future road changes.

National Land Use Database

The National Land Use Database (NLUD) is a project being engineered by the Local Government Information House, central government, the ODPM, English Partnerships (the national regeneration agency), and Britain's national mapping agency, Ordnance Survey.

The aim of the NLUD project is to provide a comprehensive and up-to-date record of vacant and derelict sites and other previously developed land and buildings that may be available for redevelopment in England.

There are two strands to the NLUD project:

- NLUD-Previously Developed Land (PDL): Collects data on vacant and derelict sites and other previously developed land and buildings that may be available for redevelopment in England.
- NLUD-Baseline: The development of a comprehensive and up-to-date land use map based on Ordnance Survey MasterMap data.

Private online information suppliers working in the residential industry

There has been a dramatic proliferation of services giving individuals or professional conveyancers the facility to conduct online searches and other related activities that can make house buying and selling easier.

The list below comprises those that have joined the umbrella organisation CoPSO (the Council of Property Search Organisations). It demonstrates two things: first, that people not hitherto linked to conveyancing or property searches have got involved (for example, private utility companies and newspaper groups); and second, that these organisations already have extensive head starts over estate agents on taking over the search functions of property transactions when, or if, the government's HIPs are introduced.

This is the list of CoPSO members. Could any estate agent have stepped in to this field in such an enterprising way?

- *Landmark Information Group* (*www.landmarkinfo.co.uk*): this supplies environmental information and large scale current and historical maps, and is owned by the *Daily Mail* newspaper.
- *OneSearch* (*www.onesearchdirect.com*): it promises a five-day turnaround of private house search data, already has over 50% of

Long waits for local government and utility company searches are already a thing of the past. Subscribers to any of the many private search firms that now exist can describe the physical area of search they want and order it online.

Source: *OneSearch*

the Scottish residential market and is expanding into England and Wales.

- *Property Search Agency* (*www.psa.co.uk*): it created Searchflow, one of the National Land Information Service's channel providers licensed by the government.
- *Property Search Group* (*www.propertysearchgroup.co.uk*): this is a Huddersfield-based company with 100 local offices through England and Wales.
- *Richards Gray* (*www.richards-gray.co.uk*): another private search firm declaring itself "the largest single employer of former local authority land charges staff in the UK".
- *Safe-Move* (*www.yorkshirewater.com*): it describes itself as "a conveyancer's one-stop shop for all the water-related and environmental information you need to help your clients make informed decisions about property purchase".
- *STL Group* (*www.stlgroup.co.uk*): a legal services firm active in property searches.
- *Groundsure* (*www.groundsure.com*): another source of information on environmental quality of land.
- *Jordans Property* (*www.jordans.co.uk*): a search and business information relating to residential and commercial property.
- *Whitbys Search Agency* (*www.personalsearch.co.uk*): a firm consisting almost entirely of former local government staff now offering searches.
- *Thames Water* (*www.twsearch.co.uk*): it offers water utility searches, not purely in Thames Water areas.
- *Severn Trent* (*www.severntrentretailsservices.co.uk*): this shrewdly offers search information on water issues — and telecom, internet and insurance services too, while you are on the site.
- *Northumbrian Water Property Solutions* (*www.nwl.co.uk*): more utility searches offered by another water body expanding into other areas.

From all of this, it is clear that the future is likely to see vastly greater amounts of online work in every aspect of selling, buying, researching and marketing property. Unwillingness to embrace this trend will simply mean that estate agents will miss out on business.

This is not to say all technological "advance" will be appropriate or even useful. For example, much debate has gone on about the veracity of residential "desktop valuations", which are beginning to be marketed in the UK.

Survey Report Management   SAVA

Property address: 52 High Street, Hemel Hempstead, Hertfordshire, HP1 3ED

Home / Navigator / Getting Started

URRN 9385

DATA INPUT:

A Introduction (Requires no input)

B General information about the property

C Other matters

D Exterior condition

E Interior condition

F Services

G Grounds, boundaries and common facilities

H Energy efficiency and recommendations

UPLOAD:

▶ Upload and view site notes and pictures

No files uploaded

REVIEW REPORTS:

▶ Review Condition Report

▶ Review Energy Report

▶ Show energy data input

▶ Finalise Report

Estate agents will be able to provide a full HIP to sellers if they wish, or buy them in from third-party suppliers that are already investing in software to create the survey and search details and the recent history of each property. Most of this will be in electronic formats.
Source: *SAVA*

Some are available on public websites. Homeowners and estate agents alike can feed in a postcode plus basic property details, such as the numbers of bedrooms and reception rooms. In return for a modest fee of under £20 paid online through a credit or debit card, a multi-page "valuation report" is returned complete with a notional figure for the property's value based on easy-to-obtain data, such as that from the Land Registry.

Some of these desktop valuations are aimed at the public (the PR people behind such schemes say the information gives sellers a good idea of what their property is worth without having to obtain the services of a professional valuer or agent; simultaneously, it allegedly informs buyers of what the "true" market value is and gives them the confidence to make appropriate offers).

In reality, the objective here is much larger scale, and it is a controversial attempt to get such desktop valuations adopted by lenders and surveying practices alike. Many property professionals who are not routinely against any form of new technology have genuine reservations about desktop valuations.

"So it doesn't matter whether a house is next to a motorway or has a view of fields? Does the same number of bedrooms mean the value is the same whether the house is 3000 or 6000 square feet? There are too many variables in real life. This won't work," warns David Peters, head of residential valuation at Knight Frank.

Chris Wood of PDQ Estates, an estate agency in the south-west of England, worked on an early version of one such service but was critical. "It gave you lots of comparables but with too little information to make sense of it all. It didn't differentiate between properties — you could easily have one comparable that was three times the price of another," he said. He described one service as "not worth the paper it's written on".

Even if homes of certain eras which were built in great volume in some suburbs of the UK can be described as "all looking like little boxes" such broad-brush desktop valuations ignore the fact the two apparently similar little boxes have very different extensions, south-facing gardens, proximity to transport links and the like. Plus, the arrival of the HCR in the HIPs shows that even the government's drive for cost-effective efficiency assumes that a personal visit is required to truly assess a property's value and condition.

But even technological mishits such as the desktop valuations now being marketed have their uses by enterprising property companies, such as My Property Value, which, in early 2005, set up *www.mypropertyvalue.co.uk*.

On the surface, this appears to be another desktop valuation site with all of its shortcomings, but the object of it is to act as a tip-off service to estate agents that someone in their area may be about to sell.

The site gives a valuation to the user who feeds in a postcode and a small number of other property details. It reckons that anyone using it is likely to want to sell — so therefore the site invites agents throughout the country to pay a subscription to the service which will then alert them if someone with a local postcode uses it, presumably prior to selling.

The site believes that the local agent can then pro-actively get in touch with the would-be seller, giving them a head start on rival agents.

My Property Value was trialled for three months in Belfast in 2004, covering three postcode districts. In response to intensive local publicity, 737 valuation enquiries were made, of which 51% were from homeowners currently selling or planning to do so within six months.

It claims that 73 sales were made as a result of agents converting the enquiries into solid instructions.

Resource Centre: Who Can Provide What?

If this book has at least encouraged you to start a "get this technology" section on your To Do list, it has been a success. It might be worthwhile checking the contents, which should fall into two categories.

Must haves:

- *Personal computers/printers/scanners*: as well as being necessary for almost all of the productivity-improving services below, a personal computer conveys a modern image, even for the single-office agency.
- *Broadband*: the faster speed of broadband may seem an unnecessary luxury until you encounter it, at which point you will never want to use the slower system again. British Telecom, in particular, offers telephone lines almost everywhere in the UK that allow stable, reliable dial-up access or cheap open-all-hours access.
- *Property management software for sales and rentals*: it makes managing thousands of buyers, sellers and properties easy and automatically produces data for websites, window cards, e-mail mailings, press advertisements and hard-copy letterheaded mailings.
- *Management information software*: this helps you manage branches, keep abreast of overheads, identify areas of savings or needs for new investment, and can profile staff productivity or the effectiveness of targeted direct mailing or press advertising.
- *Personal website*: for as little as £1500, small web design businesses will create a site for you that can then be updated easily, while

publicising its internet address on all marketing material guarantees extensive returns on very modest outlays. Also, try to include virtual tours and downloadable floor plans.

- *Professional e-mail*: e-mail is now the information channel of choice for many clients and you can contact thousands of people quickly and simultaneously for the cost of one telephone call.

- *Subscription to a portal*: over 40% of internet scouring of properties begin on portals of the Rightmove or Assertahome type — users then find out that an individual agent using the portal has a site of his own and consequently scour it. In an era of much greater mobility and in the absence of a regional or national multi-listing system, people moving from one area to another find portals invaluable.

- *Digital cameras*: for rapid production of internet and hard-copy details, these are essential, although professional photographers should still be employed for top-quality marketing of more expensive properties.

- *Smart mobile telephones or PDAs*: simple mobile telephones are no longer enough, as BlackBerry devices and PDAs can replace pen and paper notebooks, keep you in touch with e-mail and the office, and even run your business through wireless internet linking.

Nice to haves:

- *Telephone monitoring software*: it not only directs your calls to the right people but also helps staff management and gives you the feedback information to focus your advertising.

- *Laptops*: good for long-distance work and if you want to make impressive pitches after making a valuation visit, or to show a client how you can prepare a Home Information Pack.

- *Internet advertising:* try putting your agency's website details on other sites, and be part of the advertising industry's fastest-growing sector.

- *Refurbish your office*: are the logo, window displays and general image dated? Is it time to step into the 21st century?

While the PCs and printers you need will be sitting in high street shops and out-of-town superstores, where can an agent turn to for some of the more advanced forms of hardware and software we have discussed in earlier parts of the book?

This chapter ignores suppliers of the basic equipment — personal computers, laptops, mobile telephones — because they are readily available. Besides, many agents with the greatest need for injections of new technology have already got good hardware such as this ... the issue is simply that they think ownership of such kit is enough.

In these competitive days, it isn't, and agents need more customised and sophisticated equipment. When they get it, they need to upgrade it frequently.

This chapter is set out as a resource centre, in four parts. The first lists software suppliers dealing in everything from property management packages to floor plans you can read on a hand-held electronic organiser; the second is a disappointingly short list of firms majoring in specialist office hardware for agents (although many general office suppliers are likely to be well equipped). The third section looks at specialist training for agents. Although most agents say they will consider looking at IT or industry changes at a later date it is perhaps an indicator of the estate agency profession in 2005 that a list of training courses is relatively brief and, with only few exceptions, can make almost no mention of new technology or even the imminent changes in how homes are bought and sold in the UK.

Section four is a reference section on the internet. It begins with definitions of the "domain names" that exist — ".co.uk", ".com", ".org" and the like — including what they mean and what an estate agent could or could not use. It is included because, perhaps surprisingly, it was a question that came up often from estate agents consulted during the preparation of this book.

In addition, there is a list of internet property portals that exist to advertise homes for sale — or at least those portals that have survived some years of commercial uncertainty and may now be a reliable site to carry your own portfolio of homes for sale or rent.

The final internet reference is simply a recitation of the industry's only new technology guidelines — an attempt by the NAEA to set out a minimum standard for any site set up by an individual agent; it is reproduced here with the kind permission of the NAEA.

All organisations, companies and products included in this chapter are selected in a bid to provide a comprehensive directory and do not suggest endorsement of any individual, item or service.

But they will help to turn estate agents into 21st century business people, which is also the aim of this book. If it has encouraged the process in even the smallest way, it has been worthwhile.

Software providers for estate agents

Acquaint CRM — manufacturers of sales, lettings and contact management software, website designers, providers of floor plans and e-brochures. *www.acquaintcrm.co.uk*; Tel: 0118 934 4666; IMS Communications Ltd, Flagstaff House, 14 High Street, Twyford, Berkshire RG10 9AE.

Atrium Software Ltd — manufacturers of web-based software for property and asset management. *www.atriumsoft.com*; Tel: 01275 814333; Lawes House, 66–68 Bristol Road, Portishead, Bristol BS20 6QG.

Big Property Marketing — management consultancy, finance, marketing, property marketing and photography services to estate agents in London and south-east England. *www.bigpropertymarketing.co.uk*; Tel: 020 8392 3929; 20 Mortlake High Street, London SW14 8JN.

Breeze Media — manufacturers of estate agent property management software, website and corporate design and search engine optimisation software. *www.breezemedia.co.uk*; Tel: 01592 840640; Gleann Cottage, Glenlomond, Kinross, Perthshire KY13 9HF.

CFP Software — manufacturers of a range of sales, long-term and holiday lettings software. *www.cfp-software.co.uk*; Tel: 01209 313121; CFP Software Ltd, Prosper House, Cardrew Industrial Estate, Redruth, Cornwall TR15 1SS.

Core Systems — manufactures Encore Sales Management software plus SMStext messaging systems, "one-click" printing and PDA floor plan software for estate agents. *www.coresystems.co.uk*; Tel: 0870 701 4343; Unit 3, Pendeford Place, Pendeford Office Park, Wobaston Road, Wolverhampton WV9 5EL.

Deep Networks — networking specialists with experience of working on multi-branch estate agency systems. *www.deepnetworks.net*; Tel: 0800 980 5125; 9 Leigh Street, London WC1H 9EW.

Dezrez Services — manufacturers of property content management systems, SMS text message systems. *www.dezrez.com*; Tel: 01792 485588; Dezrez Services Ltd, Technium, Prince of Wales Dock, Kings Road, Swansea SA1 8PH.

DotMagic — website designers and manufacturers of website content management systems. *www.dotmagic.co.uk*; Tel: 01323 728849; 9 St Aubyns Road, Eastbourne, East Sussex BN22 7AS.

e-house — manufacturers of software for virtual tours, photography, floor plans and hard copy property brochures. *www.ehouse.co.uk*; Tel: 08700 798081; e-house, 449 High Road, Willesden, London NW10 2JJ.

Estate Agency Software Solutions — manufacturers of estate and lettings agents' software; bespoke software developers; website designers and website hosting. *www.estate-software.co.uk*; Tel: 0870 240 3596; Bullar Trees, Burnley Road East, Lumb, Rossendale, Lancashire BB4 9PQ.

Estates Today — provides internet and e-commerce software and services to commercial property industry; also provides intranet and extranet services. *www.estatestoday.co.uk*; Tel: 01483 855568.

Grosvenor Systems — manufacturers of the Propman integrated property management and accounting software. *www.grosvenorsystems.co.uk*; Tel: 020 7378 8358; Grosvenor Systems Ltd, 9 Baden Place, Crosby Row, London SE1 1YW.

Harmony Property Services — provision of floor plan, photography and property particulars services to estate and lettings agents in the Home Counties. *www.yourhomeplan.co.uk*; Tel: 01525 270 3762; Appleacres, Stoke Hammond, Milton Keynes, Buckinghamshire MK17 9ET.

Independent Software Systems Ltd — manufacturers of a range of software used by major property portals. *www.independent software.co.uk*; Tel: 01892 510100; Independent Software Systems Ltd, Unit 3, Century Place, Lamberts Road, Kent TN2 3EH.

Key-Data Systems — manufacturers and designers of property management software and property sales software applications. *www.key-data.co.uk*; Tel: 01407 811155; Key-Data House, The Plas, Llanfaelog, Ty Croes, Anglesey, Wales LL63 5TU.

MBL Digital Data — specialists in website hosting, e-commerce solutions, graphic design, web design, online applications and database integration. *www.mbl-digitalmedia.co.uk*; Tel: 0207 384 2483 and 07789 960386; MBL Digital Media, 21 Barton House, Wandsworth Bridge Road, London SW6 2PD.

Merchant Internet — internet marketing and promotion for estate agents. *www.merchantinter.net*; Tel: 0800 980 5125; 9 Leigh Street, London WC1H 9EW.

Mobile Agent — floor plan software for personal computers and hand-held PDAs. *www.themobileagent.co.uk*; Tel: 01733 558999; Prototec Software Ltd, 67a Broadway, Peterborough, Cambridgeshire PE1 1SY.

Netguides — manufacturers of software called Property Pro, a residential sales, lettings and commercial sales package. Features include property matching, virtual tours, diary and office planner, letters and advert creation. *www.netguides.co.uk/properties*; Tel: 01983 532255; Mill Cottage, Furlongs, Newport, Isle of Wight PO30 2AA.

Netstep — web designers. *www.netstep.co.uk*; Tel: 01422 200308. Netstep Corporate Communications Ltd, The Spire, Leeds Road, Lightcliffe, Halifax HX3 8NU.

Niche Software — manufactures of floor plan software for estate agents. *www.niche.co.uk*; Tel: 01242 519085; Niche Software, 144 Ryeworth Road, Cheltenham GL52 6LY.

Odyssey Interactive — manufacturers of property management software, mainly for housing associations and registered social landlords. *www.odyssey-i.com*; Tel: 0161 927 3222; Odyssey Interactive Ltd, 4th Floor, The Graftons, Stamford New Road, Altrincham, Cheshire WA14 1DQ.

ProAGENT — manufacturers of property website updating systems; website design. *www.fyldeweb.co.uk*; Tel: 01253 471557; ProAGENT, 68 North Drive, Thornton, Cleveleys, Lancashire FY5 3AH.

Property Intelligence — manufacturers of property management databases, mainly aimed at commercial sector. *www.focusnet.co.uk*; Tel: 020 7839 7684; Property Intelligence plc, Portman House, 2 Portman Street, London W1H 6EB.

Rentman Software — manufacturers of an all-inclusive residential letting and property management software package. *www.askalix.co.uk*; Tel: 0871 277 7711; 14 High Street, Wanstead, London E11 2AJ.

Sales Viewpoint.com — manufacturers of an internet-delivered property databases for sales and lettings agent; brochure and marketing options. *www.viewpoint.net.uk*; Tel: 01344 300100; VPI, Venture House, Arlington Square, Downshire Way, Bracknell, Berkshire RG12 1WA.

Shore Communications — manufacturers and sellers of telephone management software, including TIM Professional, call logging and call recording. *www.shore-comms.co.uk*; Tel: 020 8861 4070; Shore Communications Ltd, 67 Downs Road, Sutton, Surrey SM2 5NR.

Spa Microsystems — manufacturers of residential sales and lettings software. Features includes applicants and properties database, advertising, sales progression, accounts and reporting. *www.spa-micro.com*; Tel: 01344 750100; Unit 5–6, Wellington Business Park, Dukes Ride, Crowthorne RG45 6LS.

Technicweb Web Design — web designers specialising in small agencies. *www.centurion-property.co.uk*; Tel: 01376 349988; 40 Springwood Drive, Braintree, Essex CM7 2YN.

Technology Forge — manufacturers of property, asset and estate management. *www.technologyforge.com*; Tel: 01943 464844; Top Floor, Pegholme Wharfebank Business Centre, Otley, Leeds LS21 3JP.

10Ninety — manufactures a fully integrated letting management system with a search facility and office administration functions matching tenants to properties plus full landlord and tenant documentation. *www.10ninety.co.uk*; e-mail to *enquiries@10ninety.co.uk*.

Thesaurus Technology — manufacturers of estate and lettings agents' property management software. *www.housescape.co.uk*; Tel: 0870 744 2679; Thesaurus Technology, Genesys House, Sandbeck Way, Wetherby LS22 7DN.

Tri-Line Network Telephony — software sellers and manufacturers, including sales of TIM Professional, call logging and call recording. *www.tri-line.co.uk*; Tel: 020 7920 7090; Tri-Line Network telephony Ltd, 9 Printing House Yard, Shoreditch, London E2 7PR.

Vebra — manufacturers of an internet-delivered property databases for sales and lettings agent; brochure and marketing options. *www.vebra.com*; Tel: 0845 230 5333; Unit 1, Park Industrial Estate, Frogmore, St Albans AL2 2DR.

Viewpoint — manufactures software to manage sales process from initial enquiry through to final invoice, as well as creates colour electronic brochures and tracks progress of transactions. *www.viewpoint.net.uk*; Tel: 01344 300100; Venture House, Arlington Square, Downshire Way, Bracknell, Berkshire RG12 1WA.

Websky — manufactures Expert Landlord lettings agency software. *www.expertlandlord.co.uk*; Tel: 08700 771120; Websky Ltd, Tramshed Technology Centre, Beehive Yard, Walcot Street, Bath BA1 5BT.

Hardware providers for estate agents

Aspect Displays — estate agency office interior design specialists and suppliers. *www.aspectdisplays.co.uk*; Tel: 0161 763 9955; 35–36 Bury Business Centre, Kay Street, Bury, Greater Manchester BL9 6BU.

Just Displays — estate agency office interior design specialists and suppliers. *www.justdisplay.com*; Tel: 01454 411585; The Old Surgery, 12 High Street, Thornbury, Bristol BS35 2AQ.

Lane Business Systems — plasma screens, high-tech equipment for estate agency offices. *www.lanebus.com*; Tel: 020 7828 6767; Lane Business Systems, 7 Denbigh Street, London SW1V 2HF.

Leica — laser measurer manufacturers. *www.disto.com*; Tel: 01908 256500; Leica Geosystems Ltd, Davy Avenue, Knowlhill, Milton Keynes MK5 8LB.

Plasmas-Direct — flatscreen TV specialists for estate agency offices.

www.plasmas-direct.co.uk; Tel: 020 7722 0067; Visionary AV Solutions, 20 Englands Lane, Hampstead, London NW3 4TG.

Strait Line — laser measurer manufacturers. *www.straitline.com*; online purchasing.

Sonin Tools — laser measurer manufacturers. *www.sonin.com*; online purchasing.

Video Communications Systems — manufacturers of video conferenc-ing equipment. *www.videocom.co.uk*; Tel: 0870 609 2656; Video Communication Systems Ltd, PO Box 51, Saxmundham, Suffolk IP17 1ZB.

Video Conference Center — offering video-conferencing facilities within London. *www.video-conference-center.co.uk*; Tel: 020 7255 2557; 29 Harley Street, London W1G 9QR.

Video Meeting Company — this firm of video-conferencing specialists was founded by Noel Edmonds. *www.videomeetingcompany.com*; Tel: 0870 160 4466; Video Meeting Company, 3rd Floor, Beaumont House, Kensington Village, Avonmore Road, London W14 8TS.

Training for estate agents

BTMS — run by Philip Bowden, who has worked in residential and commercial estate agency, lettings, conveyancing, surveying and financial services since 1989. There are courses on:

- sales training
- management development
- executive coaching
- compliance training
- secret shopping
- recruitment
- selection.

Contact: *www.bowdentms.co.uk*; Tel: 01296 584315; or e-mail *philip.bowden@bowdentms.co.uk*.

The Darvell Group — led by former estate agent John Darvell and financial services consultant and estate agent Neil Cole. There are courses on:

- successful sales negotiation
- the sales presentation and how to sell
- generating the right business

- performance management.

Contact: *www.darvall.co.uk*

Inside Out Communications — run by estate agency marketing expert Richard Rawlings. Courses include:

- Results and Challenges in Contemporary Estate Agency (the only course currently including how agents may benefit from the Multi-Listing System and the Home Information Packs).

Contact: *www.insideoutpr.co.uk*; Tel: 01242 510504; 14 Montpellier Grove, Cheltenham, Gloucestershire GL50 2XB.

National Association of Estate Agents — training to become an estate agent involves two qualifications through the NAEA.

They are the Certificate of Practice in Estate Agency and the Certificate in Residential Lettings and Management. Study is by distance learning. Students may also work towards NVQs/SVQs in Residential Estate Agency levels 2 and 3 and Residential Property Letting and Management Agency levels 2 and 3.

General Practice Surveyors can work towards NVQ/SVQ level 4 in Valuation and/or Property Management.

NVQs/SVQs in Residential Estate Agency levels 2 and 3 and Residential Property Letting and Management Agency levels 2 and 3 are available and Foundation and Advanced Modern Apprenticeships (MAPPs) may be available for students aged between 16 and 24.

Developed by TTC Training, the NAEA describes its MAPPs programme as "ideal for young people wanting to enter the property services industry. Upon successful completion, you can achieve an NVQ level 2 or 3 in the Sale of Residential Property. It gives you the chance to earn while you learn and take the first steps towards the career of your choice. You can get a real job, you will receive training and you can obtain a qualification".

The MAPPs course covers customer service, essential law, gaining new business, the contracting process and market appraisals, selling skills, negotiating skills and progressing sales.

Other NAEA courses include:

- an introduction to successful estate agency
- essential law for residential estate agents
- the Property Misdescriptions Act and how to survive it

- successful negotiating and sales progression — increasing the number of sales you make
- conveyancing workshop: practice and procedure
- understanding property auctions — your duty of care and best advice
- effective market appraisals and profitable instructions
- working with a new home developer.

Perkins Professional Services Group — run by David Perkins, a past president of the NAEA. Courses include:

- handling complaints: in house, Ombudsman and RICS
- compliance audits
- professional training for valuers, negotiators and branches
- Property Misdescriptions Act training.

Contact: *www.davidperkins.co.uk*; Tel: 01865 310112; PO Box 176, Oxford OX2 8PD.

Royal Institution of Chartered Surveyors (RICS) — this industry giant already offers extensive training on issues as varied and arcane as arts and antiques, building control, building surveying, commercial property, construction, geomatics, hydrographic surveying, management consulting, minerals and waste, planning, quantity surveying, rural surveying and valuation as part of its commitment to continual professional development. Estate agents are actively encouraged to become RICS members.

RICS has a new programme of courses aimed at helping individuals become HIs in preparation for the HIPs being introduced in 2007. A series of road shows throughout the UK in 2005 will lead to training beginning later in the year.

Contact: *www.rics.org.uk*; Tel: 0870 333 1600; Royal Institution of Chartered Surveyors Education and Training, Surveyor Court, Westwood Way, Coventry CV4 8JE.

Surveyors and Valuers Accreditation Ltd (SAVA) — a standards setting body for surveyors, estate agents and valuers in the residential sector, grant funded by the ODPM and set up in 1999 and now an assessment centre for HI training. Courses include:

- complete Home Condition Report writing

- Home Inspector qualifications
- health and safety training
- homebuyer survey and valuation workshops
- residential building survey workshops.

Contact: *www.sava.org.uk*; Tel: 0870 837 6565; SAVA, PO Box 5603, Milton Keynes MK5 8XR.

Adam J Walker and Associates — run by management consultant and trainer Adam Walker who has written three property-related books. Courses include:

- *Introduction to Estate Agency*, for negotiators with less than one year's experience looking at how to create an outstanding impression, how to register and qualify new applicants, and how to handle enquiries, fee valuations and rude and angry people.
- *Negotiating Offers and Progressive Sales*, for branch managers and negotiators, looking at how to convert more instructions into sales, particularly overpriced or sticking properties.
- *Winning Instructions*, for branch managers, valuers, negotiators and lettings agents looking at sales presentations, converting valuations into instructions and achieving higher fee levels.
- *Managing for Success*, for proprietors, directors and branch managers, looking at the preparation of detailed business plans.

Contact: *www.adamjwalker.co.uk*; Tel: 01923 334881; Adam J Walker & Associates, Beaumont House, Nottingham Road, Heronsgate, Hertfordshire WD3 5DP.

Internet reference guide
Internet domains

Estate agents in many countries are trying to use internet domain names which differentiate them from their rivals — they want something unusual or humorous or, at the very least, different. Domain names are key to websites' online identity. It can tell people what a company is, what it does and where it is from — it is also, usually, the address for the website and e-mail to the company. Estate agents can choose whatever name they wish. Within the UK, most firms can use:

co.uk — perfect for UK-based businesses

org.uk — "org" is short for organisation, and this domain type is often used by charities and the voluntary sector

me.uk — ideal for individuals

ltd.uk — for UK limited companies, but the domain name must exactly match the registered company name

plc.uk — for UK public limited companies, where the domain name must match registered company name exactly

com — the famous one, used across the world — often thought of as American

org — an international domain for organisations (note it does not finish with ".uk")

net — originally intended for computer networks but often now used as an alternative to .com

info — a relatively new domain used by information providers and as an alternative to .com

biz — another new domain intended for use by businesses

name — the international domain for individuals is slightly different to other domains: if your name is Mike Smith, you will need to register mike.smith.name, although this may soon change

tv — originally the domain of a tiny Pacific island (Tuvalu) but now used by television programmes and companies.

Internet property portals

In addition to individual estate agents' websites there are many property portals — these are "aggregating" websites that gather property details from tens, hundreds or even thousands of agents for easy display to the public on one site. Most portals charge individual agents for this service but the limited usage information offered by portals suggest they are now the main way that most buyers access property information on the internet. Also, portals offer a broad array of allied services, such as details of removals companies, mortgage companies, neighbourhood details or market information.

Agents considering putting their portfolios on a portal need to ensure that they are likely to remain in business (many disappear after six months or a year) and that they offer reliable and respected information.

More discerning agents may also want to get information of which portals are viewed most (or "get the most hits" in the patois of the online aficionados). There is bitter rivalry between sites and most will go to great lengths to rubbish the claims of others.

www.hitwise.co.uk offers a service — at a cost — that reveals the number of hits for each of the rival property portals. Hitwise is an online measurement company that monitors sites connected with 850 businesses across the world. Founded in 1997, Hitwise is based in Melbourne but its London office daily monitors around five million home, work and educational internet users in the UK.

Its last public figures on UK residential property website hits date back to 2002, so should be treated with some caution. Interestingly, almost all of the sites still exist and are still regarded as among the market leaders:

Table 7.1 Competition Between Internet Portals

Website	Agent or portal	% market share
Vebra	Portal	14.53
Rightmove	Portal	12.78
Propertyfinder	Portal	5.01
Assertahome	Portal	4.93
Findaproperty	Portal	4.22
Your Move	Agent	4.13
Homes Online UK	Portal	2.72
Team	Agents' affinity group	2.03
Sequence	Agent	1.65
Edinburgh Property Centre	Portal	1.53
UK Property Shop	Portal	1.50
Reeds Rains	Agent	1.39
Primelocation	Portal	1.29
Homes On View	Portal	1.29
Number 1 4 Property	Portal	1.17
Good Migrations	Portal	1.06

Source: Hitwise

Aside from further reminding readers of the age of this data, some general points remain valid. First, the top half dozen sites receive substantial numbers of hits but the others quickly melt into relatively small numbers by comparison; second, portals do much better than individual agency sites.

Even many of the large corporate agents such as Bradford & Bingley (B&B), very active in the UK market at that time, do not feature here — not because they did not generate good business via the internet but because many clients found their way to B&B via a portal like Vebra or Rightmove or Assertahome.

The absence of authoritative information since that time is an indicator of the competitiveness and secrecy of the website industry, but there is little evidence that the broad principles of usage in 2002 do not hold firm three years later — except that many more members of the public now use the web to find a house.

Many of the most authoritative portals are listed below:

- *www.assertahome.co.uk* — advertises volume properties mainly at the middle and lower end of the market; the site is owned by Asserta Holdings, whose shareholders include Aviva plc and Arts Alliance; it also owns *www.propertyfinders.co.uk* and *www.gmw.co.uk*, a provider of software for estate agents and lettings agents.
- *www.findaproperty.co.uk* — specialises in volume properties across London and southern England with some international properties too; the site is owned by Hallmark Projects, which also manufactures estate agents' software.
- *www.fish4homes.co.uk* — advertises volume properties mainly at the middle and lower end of the market; the site is owned by four regional newspaper groups (Newsquest Media Group; Northcliffe Regional Newspapers Group Ltd; Trinity Mirror plc; and Guardian Media Group Regional Newspapers). *www.fish4homes.co.uk* holds a directory of estate agents, information on mortgages and conveyancing, provides neighbourhood reports, and offers a range of general advice and information on buying and selling your home.
- *www.homesandproperty.co.uk* — advertises volume properties mainly at the middle and lower end of the market; the site is owned by Associated New Media, the internet division of Associated Newspapers and their publications the *Daily Mail*, the *Mail on Sunday* and London's *Evening Standard*. It carries property portfolios of 2000 estate agents.
- *www.homesfile.co.uk* — handles sales and lettings properties, mainly in southern England.
- *www.homesonview.co.uk* — carries approximately 30,000 homes from about 500 estate agency offices.
- *www.itlhomesearch.com* — pooled site accessed from a variety of

web addresses featuring homes from various estate agents and substantial volumes of background information and links.

- *www.movewithus.co.uk* — operated by Partners in Property network of estate agents; it mostly deals with sales but some rental properties too. Also, it offers e-mail quotes for mortgages, removals, conveyancing and surveys. The owners claim the site generated £6.9 million of fees for agents in 2003.
- *www.nationalhomesnetwork.co.uk* — a grouping of 200 independent estate agents in the UK and some from overseas. A small site concentrating on the lower end of the market.
- *www.periodproperty.co.uk* — chiefly an advisory service run by enthusiasts but now with a reputation for publicising high-quality period properties on sale through estate agents.
- *www.primelocation.co.uk* — deals with homes at the middle and upper end of the market; the site is owned by a consortium of over 200 leading independent lettings and estate agents (the largest shares are held by Savills and Knight Frank). More than 650 agents list properties along with 80 house builders. Also, it carries listings of over 20,000 prime properties in 20 other countries. Primelocation also publishes a monthly magazine delivered to 110,000 pre-selected homes in prime central London.
- *www.propertyfile.co.uk* — also accessible as *www.vebra.com*, it is owned by one of the UK's leading estate agency software suppliers but also acts as a portal for agents — mostly at the middle and lower ends of the market — to advertise their portfolios.
- *www.propertyfinder.co.uk* — deals in the middle range of the market; the company is owned by Asserta Holdings.
- *www.propertylive.co.uk* — site operated by the NAEA for members to advertise their properties for sale.
- *www.rentamatic.co.uk* — this was the first "free to list" property rental website in the UK in 1999. About 70,000 properties to let, handled by letting agents and private landlords, are on the site, which claims over 150,000 hits per week.
- *www.rightmove.co.uk* — advertises volume properties mainly at the middle and lower end of the market; the site is owned by four estate agency giants (Connells; Countrywide Assured Group trading under brand names like Abbotts, Bairstow Eves, Bradford & Bingley, Bridgfords, Countrywide, Faron Sutaria, John D Wood & Co, Palmer Snell and Watson Bull & Porter; Halifax Estate Agents; and Royal & Sun Alliance). More than half of the top 100 estate agencies use *www.rightmove.co.uk* to

showcase their properties online, covering 99% of UK postcodes.
- *www.sps.net* — website for joint solicitor/estate agent firms operating in Yorkshire, Greater Manchester and Lancashire, the Midlands, Wales and North-East England.
- *www.tcpa.co.uk* — operated by The Country Property Association and displaying a small number of large country properties submitted by estate agents in southern England.
- *www.thehousehunter.co.uk* — a small site specialising in Scottish properties for sale. It is owned by House Hunter Property Marketing in Edinburgh.
- *www.themovechannel.com* — specialises in volume properties and property-related news for the UK, now with a comprehensive overseas property section; the site is owned by On The Move Ltd and prides itself as being "privately owned and wholly independent of an estate agent network, financial services institution or any other chain of businesses that could risk the integrity or objectivity of the information that is provided".

National Association of Estate Agents' internet guidelines

There has been no attempt to establish a minimum quality standard for any use of new technology by estate agents or anyone else within the residential property industry, with the exception of these guidelines prepared by the NAEA in late 2003.

Purpose and scope
In June 2001, an Internet Standards Working Party was convened by the National Association of Estate Agents, comprising invited representatives from among the leading property portals and also The Royal Institution of Chartered Surveyors.

The following guidelines that have been developed seek to identify best practice principles for the estate agency industry and the role of the property portals in doing business using a website and the internet. Conforming to these guidelines represents a significant effort on the part of the residential property sector to deliver reliable and accurate information to the consumer.

They seek to provide a minimum level of confidence for homebuyers and sellers in their usage of individual estate agents' websites and the large property portals. The standards relate to the display of information in the form of real property adverts, related products and other services connected to the process of buying and renting residential property.

In addition, the guidelines will establish improved reliability in the transfer of information between estate agents and the property portals that will give confidence to both sides in maintaining their reputation with the consumer.

Terminology
- *Applicant* — a person or persons visiting an estate agent's website or property portal with a view to buying or renting a residential property.
- *Client* — a person or persons to whom the agent owes a duty of care and for whom one is acting in return for a fee, most usually the seller, but in a growing number of cases the buyer when the agent is retained.
- *Correspondence* — any communication received by an estate agent, typically, in the form of an e-mail.
- *Estate Agent* — a business that acts for a principal in respect of transactions involving real estate.
- *User* — a person or persons using an estate agent's or property portal's website who may or may not be a client or applicant of any particular estate agent.

Transparency in estate agency as a business
- The estate agent must have a physical business address at which it can be contacted.
- The estate agent must provide easy-to-find and clearly stated information on its legally registered name or the name of its proprietors or partners, all names under which it trades, its business address, an e-mail address to which questions and complaints can be sent and a telephone number together with office hours during which that telephone line is open.
- The particulars of any accreditation to a professional body or an accreditation scheme, such as the Ombudsman for Estate Agents and/or the National Approved Letting Scheme.
- If it is part of a franchise group.
- The estate agent will ensure that, where it sub-contracts the performance of certain activities covered by these standards to a third party, sub-contracted work conforms to these standards.

Transparency solution
- A clear "Contact us" page with a link on every page of your website.
- Clear logos for all accreditations, with links to the accompanying website.
- A "Partners" page, with full contact details of partners, such as financial services, solicitors etc.

Honesty in obtaining property details
- All property adverts and details of sale will be on those properties on which the agent holds confirmed written instructions.

- Clients of properties displayed will have been notified in writing that their property is on the internet.

Honesty solutions
- A comprehensive business audit checking in and outgoing written material and your system for display/marketing of properties.

Quality in the conduct of its business
- Where a property is "under offer subject to contract" sales information will be amended accordingly.
- Where a property is no longer available for sale the agent will ensure that it is removed from display.
- Following exchange of contracts sales information must be removed or amended as sold as soon as possible.
- When the sale has completed, information must be removed from display unless it is part of collective information promoting properties sold by the agency.
- The agent will respond to enquiries and requests for relevant information received by e-mail promptly.

Quality solution
- A flexible web database that displays sales status information.
- A web database system that changes instantly as instructions change.
- An archiving system for properties.
- An automatic e-mailing system that matches clients to properties and can send standard letters and details while you work.
- A fast and reliable connection to the internet for easy e-mailing and instant incoming/outgoing e-mail.

Accuracy of property information
- All displayed property information will be accurate and up to date.
- Property particulars will be prepared to comply with The Property Misdescriptions Act 1991 and accord with the Advertising Standards Code of Practice.
- The estate agent will use reasonable endeavours to ensure that all property particulars passed on to the property portal conform to the provisions of statutory regulations relating to the sale or letting of property as from time to time may be in effect.
- The estate agent will take reasonable steps to ensure that all content passed to the property portal and consumer is of a decent and inoffensive nature.
- The estate agent will take responsibility for all accuracy and compliance issues except where the property portal has indicated in writing otherwise or has failed to honour its obligations under other parts of these guidelines in a manner which has directly led to a failure

to display accurate property information or be compliant.

- In dealings with property portals, the estate agent will take ownership for the accuracy of information sent to the property portal, whether it does so itself or via a third party, such as a software company.

Accuracy solution

Proper, printable property details online, including your header and footer, and the Property Misdescriptions Act disclaimers.

Respect for the privacy and security of estate agents' property and other users data

- The estate agent will display on its website its privacy policy, which should include:
  — what information is collected
  — the specific purpose for which the information may be used
  — whether and for what purpose the information may be disclosed to other parties
  — a contact point for Data Protection Act enquires.

- The estate agent will have available a data security policy signed by senior management describing how information is secured and procedures to ensure the policy is conformed to.
- The estate agent will ensure that information supplied by the user is stored securely, including physical security, control of access to systems, back-up and recovery of information and use of virus checkers.
- The estate agent will ensure that the user is able to change their personal details and requirements online.
- The estate agent will put in place suitable authentication processes to secure the user's access to data.
- The estate agent will ensure that the user can exclude themselves from direct mailings or other unsolicited contact by the estate agent or effected by the estate agent on behalf of others.
- The estate agent will ensure that all data is obtained, stored and managed in compliance with the Data Protection Act.
- The estate agent will undertake to keep any passwords supplied by the property portal secret and to change such passwords from time to time as requested by the property portal.
- The estate agent will ensure that all data passed to it from a property portal will be stored and managed in compliance with the Data Protection Act.

Privacy and security solution
- A cheap, effective Privacy and Data Protection policy written for your company and displayed prominently.

- A free test and audit of internal data security, anti-viral protection and back-up procedures.
- An effective web database that allows the public to log in to a secure server to change and manage their requirements.
- Linking your web property and web registration databases together.

Taking of payments from users via an estate agent's website
- Where an estate agent's website accepts payment for goods or services via its website, it should conform to and be accredited to a recognised web-accreditation service.

Payment solutions
- Connecting your website to a bank approved Payment Solutions Provider, such as Netbanx or Worldpay.

Portals
1. Transparency in its operation as a company
   The property portal must, at all times, have a physical business address at which it can be contacted. The property portal must provide information on its site about itself, its policies and its terms and conditions. The information must be easy to find and clearly stated. Information should include:

   - Its legally registered name or the name of its proprietors or partners, all names under which it trades, the address of its registered office, its physical business address, an e-mail address to which questions and complaints can be sent and a telephone number together with the hours during which that telephone line is open.
   - In the event the property portal takes payment for services from users, it should make clear the services and products offered, prices, taxes and other charges payable, payment arrangements, delivery times, any restrictions on customers served and any warranties, guarantees or other conditions, including cancellation and refund terms.
   - Information on its responsibilities and complaints and resolution procedures, including how to submit a complaint.
   - Details of its compliance with relevant legislation or regulation of its activities.

   The property portal will ensure that, where it subcontracts the performance of certain activities covered by these standards to a third party, subcontracted work conforms to these standards.

2. Honesty in the obtaining of property details

The property portal will use reasonable endeavours to ensure that all properties displayed have been obtained with the permission of the client and the agent appointed by the client.

The property portal will either enter into a direct commercial relationship with estate agents to display their properties or, in the case where another party (for example, a software supplier) has the commercial rights freely entered into by the estate agent to pass on their property details, will notify the estate agent that their properties will appear on the property portal.

The property portal will maintain and make available a register of all those organisations (and their URLs) to whom property details are passed or to whom it provides software allowing its property database to be searched. The register should be available on written request and, where it exists, on a part of the website accessible to estate agents.

3. Quality in the conduct of its business

The property portal will use reasonable endeavours to maintain access to the website for users and estate agents, including having in place site monitoring procedures and procedures for prompt restart of the site in the event of failure.

The property portal will use reasonable endeavours to ensure that adequate information is presented about individual properties to allow the user to make high-level selections of suitability in a similar manner to the presentation of information in a property newspaper.

The property portal will conform to the Advertising Standards Code of Practice in all its advertising, including any statements made on the website.

The property portal will allow its estate agent members to indicate which professional bodies and/or industry accreditation schemes they belong or subscribe to (for example, Association of Residential Lettings Agents (ARLA), National Association of Estate Agents (NAEA), Royal Institution of Chartered Surveyors (RICS), Ombudsman Scheme (OEA) and National Approved Letting Scheme (NALS), although is under no obligation to require membership of any specific body.

4. Shared responsibility for the accuracy of property information and compliance

The property portal will publish its policy on accuracy of information, indicating how information is obtained and with what frequency it is updated.

The property portal will publish its policy on responsibility for property information and compliance and also as to what codes and standards it complies with.

The property portal will use reasonable endeavours to ensure that estate agents from which it accepts properties for listing conform to the provisions of the Property Misdescriptions Act 1991 and such other regulatory provisions relating to the sale of properties as from time to time may be in effect. Reasonable endeavours would be evidenced by the existence of a contract between the portal and the estate agent specifying the estate agent's obligations or clearly published terms that the estate agent is deemed to have accepted by accessing the portal's site or site administration systems.

The property portal will provide the facilities to display the status of a property in terms of being for sale, under offer/sold subject to contract and sold.

The property portal will, if it is brought to its attention, remove duplicate details of the same property where those duplicate details have been provided by the same estate agent.

The property portal will apply changes to and remove property details based on updates supplied directly from the estate agent or from the estate agent's chosen business partner (typically a software provider) within 24 hours of their receipt.

The property portal will provide facilities for the estate agent to be able to amend and to be able to remove a property from the property portal within 24 hours, independently of any regularly applied data feeds.

The property portal will take responsibility for all property accuracy and compliance issues arising directly from a failure to honour its obligations under other parts of these standards, where such failure has directly led to a failure to display accurate property information or be compliant.

5. Clear respect for the privacy and security of estate agents' property and other data
The property portal will ensure that information supplied by the estate agent is stored securely, including physical security, control of access to systems, back-up and recovery of information and use of virus checkers.

The property portal will put in place suitable authentication processes to secure the estate agents' access to their data.

The property portal will provide contact details on the site for any estate agent wishing to contact the portal by e-mail and telephone and will undertake to respond to such contacts within one working day.

6. Clear respect for privacy and security of users' personal data
The property portal will display on the site its privacy policy, which should include:

- What information is collected.
- The specific purposes for which the information may be used.
- Whether and for what purposes the information may be disclosed to other parties.

- A contact point for Data Protection Act enquiries.

Policy on the use of cookies:
- The property portal will have available a data security policy signed by senior management describing how information is secured and procedures to ensure the policy is conformed to.
- The property portal will ensure that information supplied by the user is stored securely, including physical security, control of access to systems, back-up and recovery of information and use of virus checkers.
- The property portal will ensure that the user is able to change their personal details and requirements online.

The property portal will put in place suitable authentication processes to secure the user's access to their data

The property portal will ensure that the user can exclude themselves from direct mailings or other unsolicited contact by the property portal or effected by the property portal on behalf of others.

The property portal will ensure that all data is obtained, stored and managed in compliance with the Data Protection Act.

The estate agent will ensure that all data passed to it from a property portal will be stored and managed in compliance with the Data Protection Act.

7. Transparency in role as a communicator between user and agent
The property portal will make clear in all instances where information on a specific property is displayed, who the estate agent is and how the estate agent can be contacted.

The property portal will pass all enquiries and correspondence intended for the estate agent to the estate agent within 24 hours of receipt, or, in the event of a technical failure, will notify the estate agent within 24 hours of the existence of a technical problem.

The property portal will not use the contents of correspondence for its own commercial gain, and, specifically, undertakes not to pass details of users or their requirements provided in relation to an estate agent selling a particular property to any other agent.

8. Taking of payments from users via the website
Where a property portal accepts payments for goods and services via its website, it should conform to and be accredited by a recognised web-accreditation service.

Index